RILEM State-of-the-Art Reports

RILEM STATE-OF-THE-ART REPORTS

Volume 22

RILEM, The International Union of Laboratories and Experts in Construction Materials, Systems and Structures, founded in 1947, is a non-governmental scientific association whose goal is to contribute to progress in the construction sciences, techniques and industries, essentially by means of the communication it fosters between research and practice. RILEM's focus is on construction materials and their use in building and civil engineering structures, covering all phases of the building process from manufacture to use and recycling of materials. More information on RILEM and its previous publications can be found on www.RILEM.net.

The RILEM State-of-the-Art Reports (STAR) are produced by the Technical Committees. They represent one of the most important outputs that RILEM generates—high level scientific and engineering reports that provide cutting edge knowledge in a given field. The work of the TCs is one of RILEM's key functions.

Members of a TC are experts in their field and give their time freely to share their expertise. As a result, the broader scientific community benefits greatly from RILEM's activities.

RILEM's stated objective is to disseminate this information as widely as possible to the scientific community. RILEM therefore considers the STAR reports of its TCs as of highest importance, and encourages their publication whenever possible.

The information in this and similar reports is mostly pre-normative in the sense that it provides the underlying scientific fundamentals on which standards and codes of practice are based. Without such a solid scientific basis, construction practice will be less than efficient or economical.

It is RILEM's hope that this information will be of wide use to the scientific community.

rilem

More information about this series at http://www.springer.com/series/8780

Gideon P.A.G. van Zijl · Volker Slowik
Editors

A Framework for Durability Design with Strain-Hardening Cement-Based Composites (SHCC)

State-of-the-Art Report of the RILEM Technical Committee 240-FDS

Springer

Editors
Gideon P.A.G. van Zijl
Department of Civil Engineering
University of Stellenbosch
Matieland
South Africa

Volker Slowik
Leipzig University of Applied Sciences
Leipzig
Germany

ISSN 2213-204X ISSN 2213-2031 (electronic)
RILEM State-of-the-Art Reports
ISBN 978-94-024-1012-9 ISBN 978-94-024-1013-6 (eBook)
DOI 10.1007/978-94-024-1013-6

Library of Congress Control Number: 2016955798

Printed on acid-free paper

This Springer imprint is published by Springer Nature
The registered company is Springer Science+Business Media B.V.
The registered company address is: Van Godewijckstraat 30, 3311 GX Dordrecht, The Netherlands

RILEM Publications

The following list is presenting the global offer of RILEM Publications, sorted by series. Each publication is available in printed version and/or in online version.

RILEM PROCEEDINGS (PRO)

PRO 1: Durability of High Performance Concrete (ISBN: 2-912143-03-9); *Ed. H. Sommer*

PRO 2: Chloride Penetration into Concrete (ISBN: 2-912143-00-04); *Eds. L.-O. Nilsson and J.-P. Ollivier*

PRO 3: Evaluation and Strengthening of Existing Masonry Structures (ISBN: 2-912143-02-0); *Eds. L. Binda and C. Modena*

PRO 4: Concrete: From Material to Structure (ISBN: 2-912143-04-7); *Eds. J.-P. Bournazel and Y. Malier*

PRO 5: The Role of Admixtures in High Performance Concrete (ISBN: 2-912143-05-5); *Eds. J. G. Cabrera and R. Rivera-Villarreal*

PRO 6: High Performance Fiber Reinforced Cement Composites—HPFRCC 3 (ISBN: 2-912143-06-3); *Eds. H. W. Reinhardt and A. E. Naaman*

PRO 7: 1st International RILEM Symposium on Self-Compacting Concrete (ISBN: 2-912143-09-8); *Eds. Å. Skarendahl and Ö. Petersson*

PRO 8: International RILEM Symposium on Timber Engineering (ISBN: 2-912143-10-1); *Ed. L. Boström*

PRO 9: 2nd International RILEM Symposium on Adhesion between Polymers and Concrete ISAP '99 (ISBN: 2-912143-11-X); *Eds. Y. Ohama and M. Puterman*

PRO 10: 3rd International RILEM Symposium on Durability of Building and Construction Sealants (ISBN: 2-912143-13-6); *Eds. A. T. Wolf*

PRO 11: 4th International RILEM Conference on Reflective Cracking in Pavements (ISBN: 2-912143-14-4); *Eds. A. O. Abd El Halim, D. A. Taylor and El H. H. Mohamed*

PRO 12: International RILEM Workshop on Historic Mortars: Characteristics and Tests (ISBN: 2-912143-15-2); *Eds. P. Bartos, C. Groot and J. J. Hughes*

PRO 13: 2nd International RILEM Symposium on Hydration and Setting (ISBN: 2-912143-16-0); *Ed. A. Nonat*

PRO 14: Integrated Life-Cycle Design of Materials and Structures—ILCDES 2000 (ISBN: 951-758-408-3); (ISSN: 0356-9403); *Ed. S. Sarja*

PRO 15: Fifth RILEM Symposium on Fibre-Reinforced Concretes (FRC)—BEFIB'2000 (ISBN: 2-912143-18-7); *Eds. P. Rossi and G. Chanvillard*

PRO 16: Life Prediction and Management of Concrete Structures (ISBN: 2-912143-19-5); *Ed. D. Naus*

PRO 17: Shrinkage of Concrete—Shrinkage 2000 (ISBN: 2-912143-20-9); *Eds. V. Baroghel-Bouny and P.-C. Aïtcin*

PRO 18: Measurement and Interpretation of the On-Site Corrosion Rate (ISBN: 2-912143-21-7); *Eds. C. Andrade, C. Alonso, J. Fullea, J. Polimon and J. Rodriguez*

PRO 19: Testing and Modelling the Chloride Ingress into Concrete (ISBN: 2-912143-22-5); *Eds. C. Andrade and J. Kropp*

PRO 20: 1st International RILEM Workshop on Microbial Impacts on Building Materials (CD 02) (e-ISBN 978-2-35158-013-4); *Ed. M. Ribas Silva*

PRO 21: International RILEM Symposium on Connections between Steel and Concrete (ISBN: 2-912143-25-X); *Ed. R. Eligehausen*

PRO 22: International RILEM Symposium on Joints in Timber Structures (ISBN: 2-912143-28-4); *Eds. S. Aicher and H.-W. Reinhardt*

PRO 23: International RILEM Conference on Early Age Cracking in Cementitious Systems (ISBN: 2-912143-29-2); *Eds. K. Kovler and A. Bentur*

PRO 24: 2nd International RILEM Workshop on Frost Resistance of Concrete (ISBN: 2-912143-30-6); *Eds. M. J. Setzer, R. Auberg and H.-J. Keck*

PRO 25: International RILEM Workshop on Frost Damage in Concrete (ISBN: 2-912143-31-4); *Eds. D. J. Janssen, M. J. Setzer and M. B. Snyder*

PRO 26: International RILEM Workshop on On-Site Control and Evaluation of Masonry Structures (ISBN: 2-912143-34-9); *Eds. L. Binda and R. C. de Vekey*

PRO 27: International RILEM Symposium on Building Joint Sealants (CD03); *Ed. A. T. Wolf*

PRO 28: 6th International RILEM Symposium on Performance Testing and Evaluation of Bituminous Materials—PTEBM'03 (ISBN: 2-912143-35-7; e-ISBN: 978-2-912143-77-8); *Ed. M. N. Partl*

PRO 29: 2nd International RILEM Workshop on Life Prediction and Ageing Management of Concrete Structures (ISBN: 2-912143-36-5); *Ed. D. J. Naus*

PRO 30: 4th International RILEM Workshop on High Performance Fiber Reinforced Cement Composites—HPFRCC 4 (ISBN: 2-912143-37-3); *Eds. A. E. Naaman and H. W. Reinhardt*

PRO 31: International RILEM Workshop on Test and Design Methods for Steel Fibre Reinforced Concrete: Background and Experiences (ISBN: 2-912143-38-1); *Eds. B. Schnütgen and L. Vandewalle*

PRO 32: International Conference on Advances in Concrete and Structures 2 vol. (ISBN (set): 2-912143-41-1); *Eds. Ying-shu Yuan, Surendra P. Shah and Heng-lin Lü*

PRO 33: 3rd International Symposium on Self-Compacting Concrete (ISBN: 2-912143-42-X); *Eds. Ó. Wallevik and I. Níelsson*

PRO 34: International RILEM Conference on Microbial Impact on Building Materials (ISBN: 2-912143-43-8); *Ed. M. Ribas Silva*

PRO 35: International RILEM TC 186-ISA on Internal Sulfate Attack and Delayed Ettringite Formation (ISBN: 2-912143-44-6); *Eds. K. Scrivener and J. Skalny*

PRO 36: International RILEM Symposium on Concrete Science and Engineering—A Tribute to Arnon Bentur (ISBN: 2-912143-46-2); *Eds. K. Kovler, J. Marchand, S. Mindess and J. Weiss*

PRO 37: 5th International RILEM Conference on Cracking in Pavements—Mitigation, Risk Assessment and Prevention (ISBN: 2-912143-47-0); *Eds. C. Petit, I. Al-Qadi and A. Millien*

PRO 38: 3rd International RILEM Workshop on Testing and Modelling the Chloride Ingress into Concrete (ISBN: 2-912143-48-9); *Eds. C. Andrade and J. Kropp*

PRO 39: 6th International RILEM Symposium on Fibre-Reinforced Concretes—BEFIB 2004 (ISBN: 2-912143-51-9); *Eds. M. Di Prisco, R. Felicetti and G. A. Plizzari*

PRO 40: International RILEM Conference on the Use of Recycled Materials in Buildings and Structures (ISBN: 2-912143-52-7); *Eds. E. Vázquez, Ch. F. Hendriks and G. M. T. Janssen*

PRO 41: RILEM International Symposium on Environment-Conscious Materials and Systems for Sustainable Development (ISBN: 2-912143-55-1); *Eds. N. Kashino and Y. Ohama*

PRO 42: SCC'2005—China: 1st International Symposium on Design, Performance and Use of Self-Consolidating Concrete (ISBN: 2-912143-61-6); *Eds. Zhiwu Yu, Caijun Shi, Kamal Henri Khayat and Youjun Xie*

PRO 43: International RILEM Workshop on Bonded Concrete Overlays (e-ISBN: 2-912143-83-7); *Eds. J. L. Granju and J. Silfwerbrand*

PRO 44: 2nd International RILEM Workshop on Microbial Impacts on Building Materials (CD11) (e-ISBN: 2-912143-84-5); *Ed. M. Ribas Silva*

PRO 45: 2nd International Symposium on Nanotechnology in Construction, Bilbao (ISBN: 2-912143-87-X); *Eds. Peter J. M. Bartos, Yolanda de Miguel and Antonio Porro*

PRO 46: ConcreteLife'06—International RILEM-JCI Seminar on Concrete Durability and Service Life Planning: Curing, Crack Control, Performance in Harsh Environments (ISBN: 2-912143-89-6); *Ed. K. Kovler*

PRO 47: International RILEM Workshop on Performance Based Evaluation and Indicators for Concrete Durability (ISBN: 978-2-912143-95-2); *Eds. V. Baroghel-Bouny, C. Andrade, R. Torrent and K. Scrivener*

PRO 48: 1st International RILEM Symposium on Advances in Concrete through Science and Engineering (e-ISBN: 2-912143-92-6); *Eds. J. Weiss, K. Kovler, J. Marchand, and S. Mindess*

PRO 49: International RILEM Workshop on High Performance Fiber Reinforced Cementitious Composites in Structural Applications (ISBN: 2-912143-93-4); *Eds. G. Fischer and V.C. Li*

PRO 50: 1st International RILEM Symposium on Textile Reinforced Concrete (ISBN: 2-912143-97-7); *Eds. Josef Hegger, Wolfgang Brameshuber and Norbert Will*

PRO 51: 2nd International Symposium on Advances in Concrete through Science and Engineering (ISBN: 2-35158-003-6; e-ISBN: 2-35158-002-8); *Eds. J. Marchand, B. Bissonnette, R. Gagné, M. Jolin and F. Paradis*

PRO 52: Volume Changes of Hardening Concrete: Testing and Mitigation (ISBN: 2-35158-004-4; e-ISBN: 2-35158-005-2); *Eds. O. M. Jensen, P. Lura and K. Kovler*

PRO 53: High Performance Fiber Reinforced Cement Composites—HPFRCC5 (ISBN: 978-2-35158-046-2); *Eds. H. W. Reinhardt and A. E. Naaman*

PRO 54: 5th International RILEM Symposium on Self-Compacting Concrete (ISBN: 978-2-35158-047-9); *Eds. G. De Schutter and V. Boel*

PRO 55: International RILEM Symposium Photocatalysis, Environment and Construction Materials (ISBN: 978-2-35158-056-1); *Eds. P. Baglioni and L. Cassar*

PRO 56: International RILEM Workshop on Integral Service Life Modelling of Concrete Structures (ISBN 978-2-35158-058-5); *Eds. R. M. Ferreira, J. Gulikers and C. Andrade*

PRO 57: RILEM Workshop on Performance of cement-based materials in aggressive aqueous environments (e-ISBN: 978-2-35158-059-2); *Ed. N. De Belie*

PRO 58: International RILEM Symposium on Concrete Modelling—CONMOD'08 (ISBN: 978-2-35158-060-8); *Eds. E. Schlangen and G. De Schutter*

PRO 59: International RILEM Conference on On Site Assessment of Concrete, Masonry and Timber Structures—SACoMaTiS 2008 (ISBN set: 978-2-35158-061-5); *Eds. L. Binda, M. di Prisco and R. Felicetti*

PRO 60: Seventh RILEM International Symposium on Fibre Reinforced Concrete: Design and Applications—BEFIB 2008 (ISBN: 978-2-35158-064-6); *Ed. R. Gettu*

PRO 61: 1st International Conference on Microstructure Related Durability of Cementitious Composites 2 vol., (ISBN: 978-2-35158-065-3); *Eds. W. Sun, K. van Breugel, C. Miao, G. Ye and H. Chen*

PRO 62: NSF/RILEM Workshop: In-situ Evaluation of Historic Wood and Masonry Structures (e-ISBN: 978-2-35158-068-4); *Eds. B. Kasal, R. Anthony and M. Drdácký*

PRO 63: Concrete in Aggressive Aqueous Environments: Performance, Testing and Modelling, 2 vol., (ISBN: 978-2-35158-071-4); *Eds. M. G. Alexander and A. Bertron*

PRO 64: Long Term Performance of Cementitious Barriers and Reinforced Concrete in Nuclear Power Plants and Waste Management—NUCPERF 2009 (ISBN: 978-2-35158-072-1); *Eds. V. L'Hostis, R. Gens, C. Gallé*

PRO 65: Design Performance and Use of Self-consolidating Concrete—SCC'2009 (ISBN: 978-2-35158-073-8); *Eds. C. Shi, Z. Yu, K. H. Khayat and P. Yan*

PRO 66: 2nd International RILEM Workshop on Concrete Durability and Service Life Planning—ConcreteLife'09 (ISBN: 978-2-35158-074-5); *Ed. K. Kovler*

PRO 67: Repairs Mortars for Historic Masonry (e-ISBN: 978-2-35158-083-7); *Ed. C. Groot*

PRO 68: Proceedings of the 3rd International RILEM Symposium on 'Rheology of Cement Suspensions such as Fresh Concrete (ISBN 978-2-35158-091-2); *Eds. O. H. Wallevik, S. Kubens and S. Oesterheld*

PRO 69: 3rd International PhD Student Workshop on 'Modelling the Durability of Reinforced Concrete (ISBN: 978-2-35158-095-0); *Eds. R. M. Ferreira, J. Gulikers and C. Andrade*

PRO 70: 2nd International Conference on 'Service Life Design for Infrastructure' (ISBN set: 978-2-35158-096-7, e-ISBN: 978-2-35158-097-4); *Ed. K. van Breugel, G. Ye and Y. Yuan*

PRO 71: Advances in Civil Engineering Materials—The 50-year Teaching Anniversary of Prof. Sun Wei' (ISBN: 978-2-35158-098-1; e-ISBN: 978-2-35158-099-8); *Eds. C. Miao, G. Ye, and H. Chen*

PRO 72: First International Conference on 'Advances in Chemically-Activated Materials—CAM'2010' (2010), 264 pp, ISBN: 978-2-35158-101-8; e-ISBN: 978-2-35158-115-5, *Eds. Caijun Shi and Xiaodong Shen*

PRO 73: 2nd International Conference on 'Waste Engineering and Management—ICWEM 2010' (2010), 894 pp, ISBN: 978-2-35158-102-5; e-ISBN: 978-2-35158-103-2, *Eds. J. Zh. Xiao, Y. Zhang, M. S. Cheung and R. Chu*

PRO 74: International RILEM Conference on 'Use of Superabsorsorbent Polymers and Other New Addditives in Concrete' (2010) 374 pp., ISBN: 978-2-35158-104-9; e-ISBN: 978-2-35158-105-6; *Eds. O.M. Jensen, M.T. Hasholt, and S. Laustsen*

PRO 75: International Conference on 'Material Science—2nd ICTRC—Textile Reinforced Concrete—Theme 1' (2010) 436 pp., ISBN: 978-2-35158-106-3; e-ISBN: 978-2-35158-107-0; *Ed. W. Brameshuber*

PRO 76: International Conference on 'Material Science—HetMat—Modelling of Heterogeneous Materials—Theme 2' (2010) 255 pp., ISBN: 978-2-35158-108-7; e-ISBN: 978-2-35158-109-4; *Ed. W. Brameshuber*

PRO 77: International Conference on 'Material Science—AdIPoC—Additions Improving Properties of Concrete—Theme 3' (2010) 459 pp., ISBN: 978-2-35158-110-0; e-ISBN: 978-2-35158-111-7; *Ed. W. Brameshuber*

PRO 78: 2nd Historic Mortars Conference and RILEM TC 203-RHM Final Workshop—HMC2010 (2010) 1416 pp., e-ISBN: 978-2-35158-112-4; *Eds J. Válek, C. Groot, and J. J. Hughes*

PRO 79: International RILEM Conference on Advances in Construction Materials Through Science and Engineering (2011) 213 pp., e-ISBN: 978-2-35158-117-9; *Eds Christopher Leung and K.T. Wan*

PRO 80: 2nd International RILEM Conference on Concrete Spalling due to Fire Exposure (2011) 453 pp., ISBN: 978-2-35158-118-6, e-ISBN: 978-2-35158-119-3; *Eds E.A.B. Koenders and F. Dehn*

PRO 81: 2nd International RILEM Conference on Strain Hardening Cementitious Composites (SHCC2-Rio) (2011) 451 pp., ISBN: 978-2-35158-120-9, e-ISBN: 978-2-35158-121-6; *Eds R.D. Toledo Filho, F.A. Silva, E.A.B. Koenders and E.M.R. Fairbairn*

PRO 82: 2nd International RILEM Conference on Progress of Recycling in the Built Environment (2011) 507 pp., e-ISBN: 978-2-35158-122-3; *Eds V.M. John, E. Vazquez, S.C. Angulo and C. Ulsen*

PRO 83: 2nd International Conference on Microstructural-related Durability of Cementitious Composites (2012) 250 pp., ISBN: 978-2-35158-129-2; e-ISBN: 978-2-35158-123-0; *Eds G. Ye, K. van Breugel, W. Sun and C. Miao*

PRO 85: RILEM-JCI International Workshop on Crack Control of Mass Concrete and Related issues concerning Early-Age of Concrete Structures—ConCrack 3—Control of Cracking in Concrete Structures 3 (2012) 237 pp., ISBN: 978-2-35158-125-4; e-ISBN: 978-2-35158-126-1; *Eds F. Toutlemonde and J.-M. Torrenti*

PRO 86: International Symposium on Life Cycle Assessment and Construction (2012) 414 pp., ISBN: 978-2-35158-127-8, e-ISBN: 978-2-35158-128-5; *Eds A. Ventura and C. de la Roche*

PRO 87: UHPFRC 2013—RILEM-fib-AFGC International Symposium on Ultra-High Performance Fibre-Reinforced Concrete (2013), ISBN: 978-2-35158-130-8, e-ISBN: 978-2-35158-131-5; *Eds F. Toutlemonde*

PRO 88: 8th RILEM International Symposium on Fibre Reinforced Concrete (2012) 344 pp., ISBN: 978-2-35158-132-2, e-ISBN: 978-2-35158-133-9; *Eds Joaquim A.O. Barros*

PRO 89: RILEM International workshop on performance-based specification and control of concrete durability (2014) 678 pp, ISBN: 978-2-35158-135-3, e-ISBN: 978-2-35158-136-0; *Eds. D. Bjegović, H. Beushausen and M. Serdar*

PRO 90: 7th RILEM International Conference on Self-Compacting Concrete and of the 1st RILEM International Conference on Rheology and Processing of Construction Materials (2013) 396 pp, ISBN: 978-2-35158-137-7, e-ISBN: 978-2-35158-138-4; *Eds. Nicolas Roussel and Hela Bessaies-Bey*

PRO 91: CONMOD 2014—RILEM International Symposium on Concrete Modelling (2014), ISBN: 978-2-35158-139-1; e-ISBN: 978-2-35158-140-7; *Eds. Kefei Li, Peiyu Yan and Rongwei Yang*

PRO 92: CAM 2014—2nd International Conference on advances in chemically-activated materials (2014) 392 pp., ISBN: 978-2-35158-141-4; e-ISBN: 978-2-35158-142-1; *Eds. Caijun Shi and Xiadong Shen*

PRO 93: SCC 2014—3rd International Symposium on Design, Performance and Use of Self-Consolidating Concrete (2014) 438 pp., ISBN: 978-2-35158-143-8; e-ISBN: 978-2-35158-144-5; *Eds. Caijun Shi, Zhihua Ou, Kamal H. Khayat*

PRO 94 (online version): HPFRCC-7—7th RILEM conference on High performance fiber reinforced cement composites (2015), e-ISBN: 978-2-35158-146-9; *Eds. H.W. Reinhardt, G.J. Parra-Montesinos, H. Garrecht*

PRO 95: International RILEM Conference on Application of superabsorbent polymers and other new admixtures in concrete construction (2014), ISBN: 978-2-35158-147-6; e-ISBN: 978-2-35158-148-3; *Eds. Viktor Mechtcherine, Christof Schroefl*

PRO 96 (online version): XIII DBMC: XIII International Conference on Durability of Building Materials and Components(2015), e-ISBN: 978-2-35158-149-0; *Eds. M. Quattrone, V.M. John*

PRO 97: SHCC3—3rd International RILEM Conference on Strain Hardening Cementitious Composites (2014), ISBN: 978-2-35158-150-6; e-ISBN: 978-2-35158-151-3; *Eds. E. Schlangen, M.G. Sierra Beltran, M. Lukovic, G. Ye*

PRO 98: FERRO-11—11th International Symposium on Ferrocement and 3rd ICTRC—International Conference on Textile Reinforced Concrete (2015), ISBN: 978-2-35158-152-0; e-ISBN: 978-2-35158-153-7; *Ed. W. Brameshuber*

PRO 99 (online version): ICBBM 2015—1st International Conference on Bio-Based Building Materials (2015), e-ISBN: 978-2-35158-154-4; *Eds. S. Amziane, M. Sonebi*

PRO 100: SCC16—RILEM Self-Consolidating Concrete Conference (2016), ISBN: 978-2-35158-156-8; e-ISBN: 978-2-35158-157-5; *Ed. Kamal H. Kayat*

PRO 101 (online version): III Progress of Recycling in the Built Environment (2015), e-ISBN: 978-2-35158-158-2; *Eds I. Martins, C. Ulsen and S. C. Angulo*

PRO 102 (online version): RILEM Conference on Microorganisms-Cementitious Materials Interactions (2016), e-ISBN: 978-2-35158-160-5; *Eds. Alexandra Bertron, Henk Jonkers, Virginie Wiktor*

PRO 103 (online version): ACESC'16—Advances in Civil Engineering and Sustainable Construction (2016), e-ISBN: 978-2-35158-161-2

PRO 104 (online version): SSCS'2015—Numerical Modeling—Strategies for Sustainable Concrete Structures (2015), e-ISBN: 978-2-35158-162-9

PRO 105: 1st International Conference on UHPC Materials and Structures (2016), ISBN: 978-2-35158-164-3, e-ISBN: 978-2-35158-165-0

PRO 106: AFGC-ACI-fib-RILEM International Conference on Ultra-High-Performance Fibre-Reinforced Concrete—UHPFRC 2017 (2017), ISBN: 978-2-35158-166-7, e-ISBN: 978-2-35158-167-4; *Eds. François Toutlemonde and Jacques Resplendino*

PRO 107 (online version): XIV DBMC—14th International Conference on Durability of Building Materials and Components (2017), e-ISBN:

978-2-35158-159-9; *Eds. Geert De Schutter, Nele De Belie, Arnold Janssens, Nathan Van Den Bossche*

PRO 108: MSSCE 2016—Innovation of Teaching in Materials and Structures (2016), ISBN: 978-2-35158-178-0, e-ISBN: 978-2-35158-179-7; *Ed. Per Goltermann*

PRO 109 (two volumes): MSSCE 2016—Service Life of Cement-Based Materials and Structures (2016), ISBN Vol. 1: 978-2-35158-170-4, Vol. 2: 978-2-35158-171-4, Set Vol. 1&2: 978-2-35158-172-8, e-ISBN : 978-2-35158-173-5; *Eds. Miguel Azenha, Ivan Gabrijel, Dirk Schlicke, Terje Kanstad and Ole Mejlhede Jensen*

PRO 110: MSSCE 2016—Historical Masonry (2016), ISBN: 978-2-35158-178-0, e-ISBN: 978-2-35158-179-7; *Eds. Inge Rörig-Dalgaard and Ioannis Ioannou*

PRO 111: MSSCE 2016—Electrochemistry in Civil Engineering (2016), ISBN: 978-2-35158-176-6, e-ISBN: 978-2-35158-177-3; *Ed. Lisbeth M. Ottosen*

PRO 112: MSSCE 2016—Moisture in Materials and Structures (2016), ISBN: 978-2-35158-178-0, e-ISBN: 978-2-35158-179-7; *Eds. Kurt Kielsgaard Hansen, Carsten Rode and Lars-Olof Nilsson*

PRO 113: MSSCE 2016—Concrete with Supplementary Cementitious Materials (2016), ISBN: 978-2-35158-178-0, e-ISBN: 978-2-35158-179-7; *Eds. Ole Mejlhede Jensen, Konstantin Kovler and Nele De Belie*

PRO 114: MSSCE 2016—Frost Action in Concrete (2016), ISBN: 978-2-35158-182-7, e-ISBN: 978-2-35158-183-4; *Eds. Marianne Tange Hasholt, Katja Fridh and R. Doug Hooton*

PRO 115: MSSCE 2016—Fresh Concrete (2016), ISBN: 978-2-35158-184-1, e-ISBN: 978-2-35158-185-8; *Eds. Lars N. Thrane, Claus Pade, Oldrich Svec and Nicolas Roussel*

PRO 116: BEFIB 2016—9th RILEM International Symposium on Fiber Reinforced Concrete (2016), ISBN: 978-2-35158-187-2, e-ISBN: 978-2-35158-186-5

PRO 117: 3rd International RILEM Conference on Microstructure Related Durability of Cementitious Composites (2016), ISBN: 978-2-35158-188-9, e-ISBN: 978-2-35158-189-6; *Eds. Changwen Miao, Wei Sun, Jiaping Liu, Huisu Chen, Guang Ye and Klaas van Breugel*

RILEM REPORTS (REP)

Report 19: Considerations for Use in Managing the Aging of Nuclear Power Plant Concrete Structures (ISBN: 2-912143-07-1); *Ed. D. J. Naus*

Report 20: Engineering and Transport Properties of the Interfacial Transition Zone in Cementitious Composites (ISBN: 2-912143-08-X); *Eds. M. G. Alexander, G. Arliguie, G. Ballivy, A. Bentur and J. Marchand*

Report 21: Durability of Building Sealants (ISBN: 2-912143-12-8); *Ed. A. T. Wolf*

Report 22: Sustainable Raw Materials—Construction and Demolition Waste (ISBN: 2-912143-17-9); *Eds. C. F. Hendriks and H. S. Pietersen*

Report 23: Self-Compacting Concrete State-of-the-Art Report (ISBN: 2-912143-23-3); *Eds. Å. Skarendahl and Ö. Petersson*

Report 24: Workability and Rheology of Fresh Concrete: Compendium of Tests (ISBN: 2-912143-32-2); *Eds. P. J. M. Bartos, M. Sonebi and A. K. Tamimi*

Report 25: Early Age Cracking in Cementitious Systems (ISBN: 2-912143-33-0); *Ed. A. Bentur*

Report 26: Towards Sustainable Roofing (Joint Committee CIB/RILEM) (CD 07) (e-ISBN 978-2-912143-65-5); *Eds. Thomas W. Hutchinson and Keith Roberts*

Report 27: Condition Assessment of Roofs (Joint Committee CIB/RILEM) (CD 08) (e-ISBN 978-2-912143-66-2); *Ed. CIB W 83/RILEM TC166-RMS*

Report 28: Final report of RILEM TC 167-COM 'Characterisation of Old Mortars with Respect to Their Repair' (ISBN: 978-2-912143-56-3); *Eds. C. Groot, G. Ashall and J. Hughes*

Report 29: Pavement Performance Prediction and Evaluation (PPPE): Interlaboratory Tests (e-ISBN: 2-912143-68-3); *Eds. M. Partl and H. Piber*

Report 30: Final Report of RILEM TC 198-URM 'Use of Recycled Materials' (ISBN: 2-912143-82-9; e-ISBN: 2-912143-69-1); *Eds. Ch. F. Hendriks, G. M. T. Janssen and E. Vázquez*

Report 31: Final Report of RILEM TC 185-ATC 'Advanced testing of cement-based materials during setting and hardening' (ISBN: 2-912143-81-0; e-ISBN: 2-912143-70-5); *Eds. H. W. Reinhardt and C. U. Grosse*

Report 32: Probabilistic Assessment of Existing Structures. A JCSS publication (ISBN 2-912143-24-1); *Ed. D. Diamantidis*

Report 33: State-of-the-Art Report of RILEM Technical Committee TC 184-IFE 'Industrial Floors' (ISBN 2-35158-006-0); *Ed. P. Seidler*

Report 34: Report of RILEM Technical Committee TC 147-FMB 'Fracture mechanics applications to anchorage and bond' Tension of Reinforced Concrete Prisms—Round Robin Analysis and Tests on Bond (e-ISBN 2-912143-91-8); *Eds. L. Elfgren and K. Noghabai*

Report 35: Final Report of RILEM Technical Committee TC 188-CSC 'Casting of Self Compacting Concrete' (ISBN 2-35158-001-X; e-ISBN: 2-912143-98-5); *Eds. Å. Skarendahl and P. Billberg*

Report 36: State-of-the-Art Report of RILEM Technical Committee TC 201-TRC 'Textile Reinforced Concrete' (ISBN 2-912143-99-3); *Ed. W. Brameshuber*

Report 37: State-of-the-Art Report of RILEM Technical Committee TC 192-ECM 'Environment-conscious construction materials and systems' (ISBN: 978-2-35158-053-0); *Eds. N. Kashino, D. Van Gemert and K. Imamoto*

Report 38: State-of-the-Art Report of RILEM Technical Committee TC 205-DSC 'Durability of Self-Compacting Concrete' (ISBN: 978-2-35158-048-6); *Eds. G. De Schutter and K. Audenaert*

Report 39: Final Report of RILEM Technical Committee TC 187-SOC 'Experimental determination of the stress-crack opening curve for concrete in tension' (ISBN 978-2-35158-049-3); *Ed. J. Planas*

Report 40: State-of-the-Art Report of RILEM Technical Committee TC 189-NEC 'Non-Destructive Evaluation of the Penetrability and Thickness of the Concrete Cover' (ISBN 978-2-35158-054-7); *Eds. R. Torrent and L. Fernández Luco*

Report 41: State-of-the-Art Report of RILEM Technical Committee TC 196-ICC 'Internal Curing of Concrete' (ISBN 978-2-35158-009-7); *Eds. K. Kovler and O. M. Jensen*

Report 42: 'Acoustic Emission and Related Non-destructive Evaluation Techniques for Crack Detection and Damage Evaluation in Concrete'—Final Report of RILEM Technical Committee 212-ACD (e-ISBN: 978-2-35158-100-1); *Ed. M. Ohtsu*

Report 45: Repair Mortars for Historic Masonry——State-of-the-Art Report of RILEM Technical Committee TC 203-RHM (e-ISBN: 978-2-35158-163-6); *Eds. Paul Maurenbrecher and Caspar Groot*

Members of RILEM TC 240-FDS

Gideon P.A.G. van Zijl, Stellenbosch, South Africa—Chairman
Flavio de Andrade Silva, Rio de Janeiro, Brazil—Secretary
Frank Altmann, Dresden, Germany
Lupita S. Beltran, Delft, The Netherlands
W.P. (Billy) Boshoff, Stellenbosch, South Africa
Esmaeel Esmaeeli, Porto, Portugal
Eduardo Fairbairn, Rio de Janeiro, Brazil
Liberato Ferrara, Milano, Italy
Gregor Fischer, Copenhagen, Denmark
Hideki Hoshiro, Tokyo, Japan
Petr Kabele, Prague, Czech Republic
Tetsushi Kanda, Tokyo, Japan
Koichi Kobayashi, Gifu, Japan
Victor C. Li, Ann Arbor, USA
Mladena Luković, Delft, The Netherlands
Viktor Mechtcherine, Dresden, Germany
Barzin Mobasher, Tempe, USA
Jiri Nemecek, Prague, Czech Republic
Claudia Ostertag, Berkeley, USA
Suvash Chandra Paul, Stellenbosch, South Africa
Alva Peled, Beersheba, Israel
Keitetsu Rokugo, Gifu, Japan
Mustafa Şahmaran, Istanbul, Turkey
Erik Schlangen, Delft, The Netherlands
Faiz U.A. Shaikh, Perth, Australia
Volker Slowik, Leipzig, Germany
Zhao Tie-jun, Qingdao, China
Romildo D. Toledo Filho, Rio de Janeiro, Brazil
Christian Wagner, Leipzig, Germany
Penggang Wang, Qingdao, China
Jan Wastiels, Brussels, Belgium
Folker H. Wittmann, Freiburg, Germany

Preface

This book captures the state of the art of the durability of strain-hardening cement-based composites (SHCC) and the durability of structures or structural elements that are manufactured in full or in part with this class of modern construction materials. It has been compiled by the RILEM Technical Committee (TC) 240-FDS: A Framework for Durability Design with SHCC. This TC was active from 2010 to 2015. The TC proposal was developed in November 2009 in Stellenbosch, South Africa, on the occasion of the International Conference on Advances in Cement-based Materials (ACM 2009), hosted by Gideon P.A.G. van Zijl. After acceptance by RILEM in March 2010, yearly meetings were held in Dresden, Germany, hosted by Viktor Mechtcherine (September 2010); in Rio de Janeiro, Brazil, at SHCC-2 hosted by Romildo Toledo Filho (November 2011); in Guimarães, Portugal, at BEFIB (September 2012); in Toledo, Spain, at FRAMCOS-8 (March 2013); in Paris, France, at the RILEM Week (September 2013); in Delft, The Netherlands, at SHCC-3 hosted by Erik Schlangen (November 2014); in Stuttgart, Germany, at HPFRCC-7 (June 2015); and in Leipzig, Germany (October 2015).

The book is subdivided into ten chapters, authored by various members:

Chapter 1 Introduction: Crack Distribution and Durability of SHCC, by Gideon P.A.G. van Zijl, William P. Boshoff, Christian Wagner and Volker Slowik.

Chapter 2 Transfer of Fluids, Gases and Ions in and Through Cracked and Uncracked Composites, by Christian Wagner, Volker Slowik, Gideon P.A.G. van Zijl, William P. Boshoff, Suvash C. Paul, Viktor Mechtcherine and Koichi Kobayashi.

Chapter 3 Fibre Durability, by Flavio A. Silva, Alva Peled, Bartosz Zukowski and Romildo D. Toledo Filho.

Chapter 4 Chemical Processes, by Erik Schlangen, Gideon P.A.G. van Zijl and Petr Kabele.

Chapter 5 Influence of Low Temperatures, by Koichi Kobayashi and Folker H. Wittmann.

Chapter 6 Influence of Elevated Temperatures, by Flavio A. Silva, Barzin Mobasher, Alva Peled, Dimas A.S. Rambo and Romildo D. Toledo Filho.

Chapter 7 Abrasion, by Volker Slowik, Steffen Müller, Christian Wagner and Viktor Mechtcherine.

Chapter 8 Behaviour of Bonded SHCC Overlay Systems, by Volker Slowik, Mladena Luković, Christian Wagner and Gideon P.A.G. van Zijl.

Chapter 9 Reinforcing Bar Corrosion, by Koichi Kobayashi, Suvash C. Paul and Gideon P.A.G. van Zijl.

Chapter 10 Durability and Service Life Design Concepts for Structures and (Non-)Structural Members Made of or Strengthened/Repaired with SHCC, by Viktor Mechtcherine, Frank Altmann and Gideon P.A.G. van Zijl.

A process of review was followed. Independent experts were appointed to review each chapter. We gratefully acknowledge the review performed by Prof. Flavio A. Silva (Chap. 1), Dr. Mladena Luković (Chaps. 2 and 10), Prof. Gideon P. A.G. van Zijl (Chaps. 3, 6 and 7), Prof. Hirozo Mihashi (Chap. 4), Prof. Volker Slowik (Chap. 5), Prof. Viktor Mechtcherine (Chap. 8), and Prof. Faiz Uddin Shaikh (Chap. 9).

An important development during the life of this TC 240-FDS is the SHCC series of conferences. The *International RILEM Workshop on High Performance Fiber Reinforced Cementitious Composites in Structural Applications*—hosted in Hawaii by Prof. Gregor Fischer in May 2005—and the *International Conference on Advances in Cement-based Materials (ACM 2009)*—hosted in Stellenbosch, South Africa, by Prof. Gideon P.A.G. van Zijl in November 2009—were followed by SHCC-2—hosted in Rio de Janeiro by Prof. Romildo D. Toledo Filho in 2011—and SHCC-3—hosted in Dordrecht, The Netherlands, by Prof. Erik Schlangen in November 2014. Following on from this groundwork, a three-year cycle is envisaged, and SHCC-4 will be held in Dresden, Germany, in 2017—hosted by Prof. Viktor Mechtcherine and co-chaired by Profs. Volker Slowik and Petr Kabele.

Finally, informed by this State-of-the-Art Report, the appropriate application of SHCC as construction material is encouraged to enhance structural durability and overall lifespan performance. By means of the SHCC conference series, continuation of research on this class of materials is envisaged. Field application and performance monitoring are considered imperatives for the validation of structural and durability design models and guidelines.

Stellenbosch, South Africa
Leipzig, Germany
July 2016

Gideon P.A.G. van Zijl
Volker Slowik

Contents

Chapter 1
Introduction: Crack Distribution and Durability of SHCC

Gideon P.A.G. van Zijl, William P. Boshoff, Christian Wagner and Volker Slowik

Abstract Inherent crack control to fine widths in strain-hardening cement-based composites (SHCC) suggests that structural elements produced from SHCC or steel-reinforced SHCC (R/SHCC) may be rendered durable by limiting the ingress rates of potentially deleterious substances. Recently, it has been reported that, while the average crack width in SHCC is maintained up to large tensile strains in excess of 3%, the maximum crack width may equal or exceed those are considered to be limiting in terms of durability. Also, the typical range in SHCC average crack width, from 50 to 100 μm, has been shown to be a threshold in water permeability, at which width permeability is restricted to several orders lower than that expected for crack widths ranging from 0.2 to 0.3 mm — a typical reinforced concrete crack width limit in durability standards. However, it has recently been shown that capillary absorption in dry, pre-cracked SHCC is a quick process, with water penetrating into fine cracks within minutes of exposure. In addition to describing these findings, this chapter sets the scene for later chapters on improved ingress rate characterisation and the actual deterioration of cracked SHCC or R/SHCC. Guidelines for the pre-cracking of SHCC towards durability testing are derived, based on the results of recent comparative testing. These include the specimen shape, size, test set-up, crack measurement to sufficient resolution, and crack width distribution presentation. Finally, the field performance of repairs, structures and structural elements produced from SHCC and R/SHCC in the past decade is reported.

Keywords Crack width · Capillary absorption · Digital image correlation · Durability

G.P.A.G. van Zijl (✉) · W.P. Boshoff
Stellenbosch University, Stellenbosch, South Africa
e-mail: gvanzijl@sun.ac.za

C. Wagner · V. Slowik
Leipzig University of Applied Sciences, Leipzig, Germany

G.P.A.G. van Zijl and V. Slowik (eds.), *A Framework for Durability Design with Strain-Hardening Cement-Based Composites (SHCC)*, RILEM State-of-the-Art Reports 22,
DOI 10.1007/978-94-024-1013-6_1

1.1 Introduction

Fibre-reinforced strain-hardening cement-based composites (SHCC) are designed to form multiple cracks, controlled to small widths and narrow spacing. The strain-hardening nature stems from the formation of successive cracks at higher resistance — even in a uniform tensile deformation field. Typical SHCC stress–strain responses and crack width distributions at various average tensile strain levels are shown in Fig. 1.1. Whereas multiple cracks are associated with high energy dissipation, for which SHCC has been exploited in, for instance, enhancing the earthquake resistance of buildings, or retrofitting in order to enhance seismic resistance, a major benefit of this class of material may lie in improved durability through crack width control and associated crack spacing.

The concept of crack width limitation for durability design is well established in reinforced concrete (RC). Design standards limit the crack width for different environments to assure the durability of RC structures to be built in these various environments. A defining characteristic of SHCC containing an embedded reinforcing steel bar, namely steel-reinforced SHCC (R/SHCC), is compatible deformation between the matrix and the bar once cracks arise. In Fig. 1.2, typical localisation can be observed in RC, leading to debonding along the bar-matrix interface in the localisation region. In contrast, the formation of multiple cracks in R/SHCC avoids such localisation and damage in the interfacial zone, which is a feature that is postulated to have significant durability benefits.

The durability of SHCC was studied by Sub-committee 2 of the RILEM (the International Union of Laboratories and Experts in Construction Materials, Systems, and Structures) Technical Committee 208-HFC in the period from 2005 to

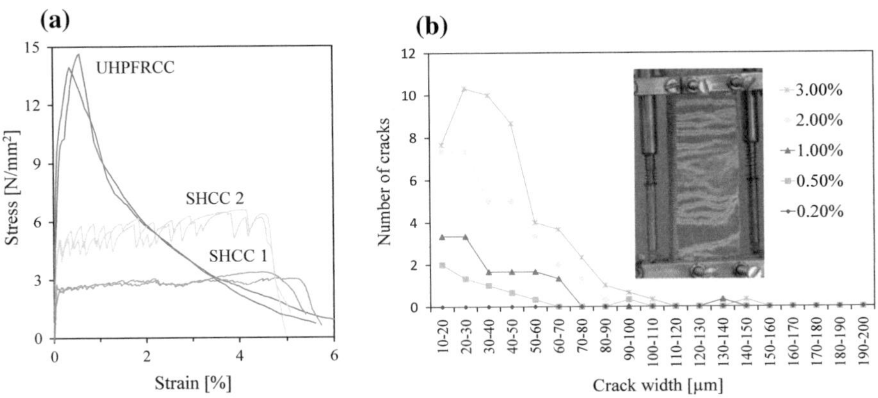

Fig. 1.1 **a** Typical tensile responses of SHCC, and **b** crack width distribution in a particular SHCC at various average tensile strain levels (reproduced from Adendorff et al. 2010)

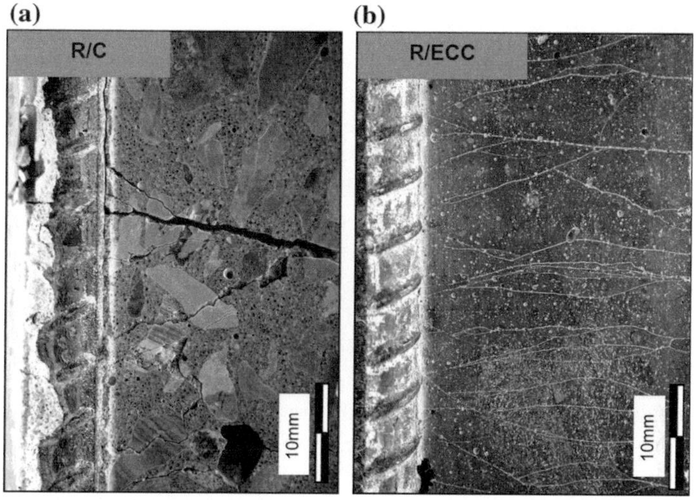

Fig. 1.2 **a** Non-uniform deformation in RC versus **b** uniform deformation in R/SHCC after crack formation, illustrating strain compatibility in R/SHCC (Fischer and Li 2002)

2010, leading to the publication of the State-of-the-Art Report (STAR) on the durability of SHCC (Van Zijl and Wittmann 2011). In addition to describing the cracking and mechanical resistance properties of SHCC, the current STAR presents a comprehensive summary of research results on durability under chemical loads, thermal loads and combined loads, on aspects related to the durability of fibres, structural elements and structures, and on life cycle considerations. This work was continued in the RILEM Technical Committee 240-FDS: Framework for Durability design with SHCC, in the period from 2010 to 2015. The core of active members of TC 208-HFC SC2 continued as members of TC 240-FDS, ensuring direct continuation of active physical laboratory experiments and particular attention to identified research needs. The summative objective was to characterise multiple crack formation and durability, and, finally, to formulate a framework for durability design with SHCC.

A particular activity started by TC 208-HFC SC2 was a comparative test series, conducted in five participating laboratories, aiming to establish the most appropriate tensile specimen size, test set-up, test procedure, and crack measurement methods. The work was continued in the lifespan of TC 240-FDS and culminated in the publication of a final paper co-authored by the five participating laboratories (Van Zijl et al. 2016). The intention was that SHCC specimens intended for durability studies in the cracked state should be consistently pre-cracked, and that the crack patterns should also be consistently measured and represented. A description of crack measurement methods for the purpose of crack characterisation towards durability is included in the current chapter as Sect. 1.6.

1.2 Scope of the Present STAR

The SHCC materials included in the present STAR are composites that contain short, discontinuous fibres and that exhibit strain-hardening tensile response to a minimum ultimate tensile strain ranging from 0.5 to 1% before localisation and subsequent strain softening. Such materials are used to ensure that the durability of the SHCC materials, which is the main focus of the present STAR, can be assured by means of appropriately characterised, reliable crack width limitation, so as to restrict the rates of ingress, as alluded to in Sect. 1.3.2 in terms of the operating strain regime. The restriction in question typically requires that a strain capacity margin exists in terms of which the likelihood that the durability threshold crack width may be exceeded is sufficiently small.

Other materials that may comply with the ultimate tensile strain limit stated above include textile-reinforced concrete (TRC), or mortars. However, the continuous fibres in layers, as well as particular fabrication processes are argued potentially to alter the durability characteristics involved, despite the apparent agreement with short-fibre SHCC in fine crack patterns being maintained to significant tensile strain levels. The above does not imply that the durability principles are not applicable to TRC, but that they have not been tested, confirmed and reported by the RILEM TC 240-FDS, although TC members, including Mechtcherine and Lieboldt (2011), Dvorkin et al. (2013), Silva et al. (2014), have researched and published details regarding the different aspects of the durability behaviour of TRC.

Emphasis in the current chapter is on crack formation in SHCC under uniaxial tension. However, applications referred to in Sect. 1.5 and the durability testing that is described in subsequent chapters include consideration of flexural cracking. In addition, SHCC may be used in combination with reinforcing steel bars (R/SHCC). As seen in Fig. 1.2, reinforcing bars may influence the crack patterns through the disturbance caused by the ribs on deformed bars and the cover depth, as reported by Paul and Van Zijl (2013). In Chap. 9, recent corrosion studies are reported on cracked R/SHCC, both unloaded and when under sustained flexural load.

1.3 Major Findings of STAR TC 208-HFC SC2

Apart from the comprehensive recording of SHCC durability, research findings of the past one-and-a-half to two decades, which are of particular significance for the further characterisation and consideration of SHCC, were identified in the work of TC 208-HFC SC2. These findings relate to the crack patterns themselves, and to the implications that such findings could have for the ingress of deleterious matter and associated deterioration processes. A second issue of importance in this respect is the potentially dominant capillary absorption into fine cracks in SHCC.

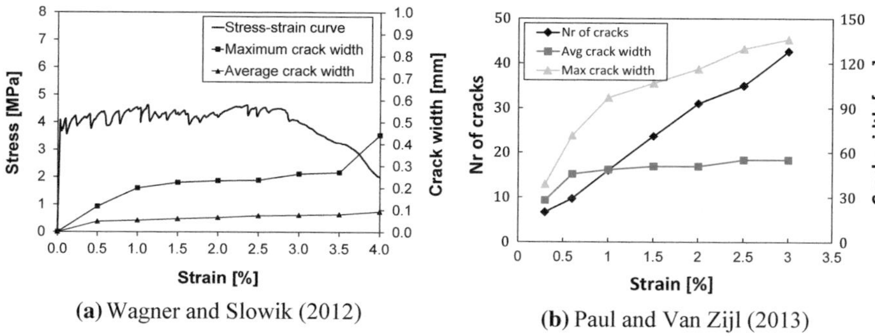

Fig. 1.3 Average and maximum crack widths in SHCC under uniaxial tension

1.3.1 Average versus Maximum Crack Width

By careful characterisation of all cracks in the gauge regions of specimens, as illustrated in Fig. 1.1b, wider cracks have been found to be several fold times the average crack width (Van Zijl and Wittmann 2010; Van Zijl et al. 2016). Whereas the average crack width remains low and reasonably constant over a large range in terms of tensile deformation, the maximum crack width may be significant and increase at an increased rate. Particular cases of such are shown in Fig. 1.3. The maximum crack width may exceed the threshold level for increased ingress rate, for instance in the case of the apparent threshold of 0.05–0.1 mm beyond which a significantly increased water permeability coefficient has been found to exist by several authors, both for concrete (Wang et al. 1997) and for SHCC (Lepech and Li 2009a). Consideration of the relatively wide cracks that occur in durability performance under exposure to different environments is of importance in terms of durability estimation and modelling.

1.3.2 Crack Formation and Capillary Absorption

In several series of comparative physical tests, SHCC has been shown to exhibit superior durability, due to the retained mechanical load resistance that is experienced under imposed and environmentally induced strain, and/or fine crack widths and associated reduced rate of ingress compared to those of mortars or concrete with localised wide cracks (Van Zijl and Wittmann 2011). For example, water permeability in cracked concrete (Wang et al. 1997) or in SHCC (Lepech and Li 2009a) increases by five to six orders of magnitude when the crack width is 0.3 mm, compared to 0.05 mm. A threshold crack width appears to exist in the region of 0.05–0.1 mm. In Fig. 1.3a, b it can be seen that the average crack width for such SHCC types remains below this threshold up to an average tensile strain of

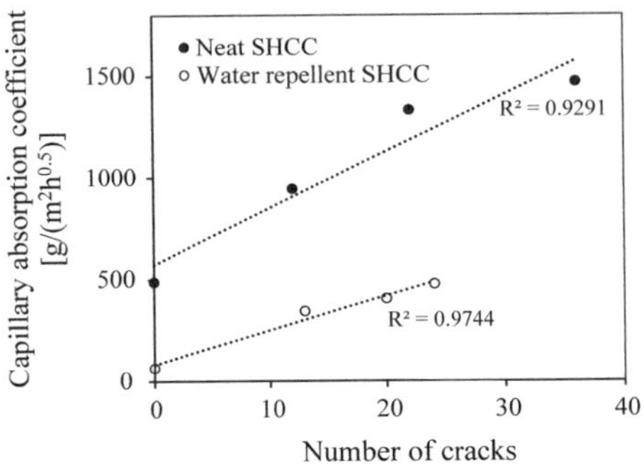

Tensile strain (%)	Neat SHCC		Water-repellent SHCC	
	No. of cracks	Avg. crack width (μm)	No. of cracks	Avg. crack width (μm)
0.5	12	50.9	13	52.3
1	22	55.2	20	61.7
2	36	55.6	24	64.4

Fig. 1.4 Capillary absorption in SHCC cracked under uniaxial tension (reworked from Wittmann et al. 2010)

about 3%. An example of enhanced durability by means of retained resistance, despite a significant amount of induced tensile strain, is the spalling resistance of cover to corroding reinforcing steel bars in the case of SHCC (Kabele et al. 1999; Lim 1996).

In dry or partially saturated SHCC, capillary absorption may, however, become a dominant mechanism of ingress into fine cracks, providing fast access to water and potentially harmful solvents. Such an ingress rate is high, with water penetrating deep into cracks within a minute of exposure to capillary absorption (Zhang et al. 2010). From Fig. 1.4, it appears that the ingress rate by the mechanism concerned increases approximately linearly with the number of cracks present, with the rate in question observed for direct tensile strain levels of 0, 0.5, 1 and 2% in the affected SHCC specimens. It must be kept in mind that the finding holds true for the early stages of capillary absorption in SHCC. The influence of the number of cracks reduces with the increasing duration of the test when the small spaces between cracks are quickly saturated. Note also, in Fig. 1.3b, that the number of cracks is roughly proportional to the tensile strain level in the strain-hardening regime. The ultimate tensile strain from a stress–strain diagram for a particular SHCC has been suggested as not being a useful parameter for durability design. Instead, a reduced strain limit should be selected that adequately limits the increased ingress of water (Wittmann et al. 2010).

1.4 Crack Formation and Durability

In order to develop a framework for durability design, and to characterise deterioration models for appropriate analysis towards reliable design, adequate sets of data for durability behaviour under various exposure types and degrees that structures may encounter are required. Obtaining such datasets typically requires durability tests to be conducted on pre-cracked specimens, or the exposure of appropriate specimens to aggressive environments or simulated deteriorating conditions during mechanical loading, so as to emulate combined action to which structures are likely to be exposed during their lifespan.

Clearly, systematic approaches are required to create a common basis for the characterisation of durability resistance. While standardised test procedures have not yet been determined, systematic approaches to the pre-cracking process, measurement of crack widths and the representation of crack pattern result data should assist in the common interpretation and understanding of subsequent durability tests towards the derivation of durability design guidelines.

1.4.1 Tensile Test Method

Sub-committee 1 of TC 208-HFC Mechanical characterisation and testing of SHCC addressed test methods to determine whether fibre-reinforced cement-based composite could be classified as SHCC, as well as methods used for recording their characteristics. The uniaxial tensile test is generally accepted to be the most direct and sound procedure for determining the tensile stress–strain response. The associated action exploits the dominant features of SHCC in field application.

Mechtcherine (2007) and Mechtcherine and Jun (2008) present important and illustrative background to the SHCC tensile specimen shape, the casting direction, the age and the curing regime, as well as the boundary conditions in the testing machine, and in the type of test control used. The stress–deformation responses of different specimen shapes in the categories (i) dumbbell-shaped, and (ii) coupon, or prismatic, plate type are compared, also showing the influence of non-rotational and rotational fixity in the test machines of the specimen ends in relation to the response, as is illustrated in Fig. 1.5.

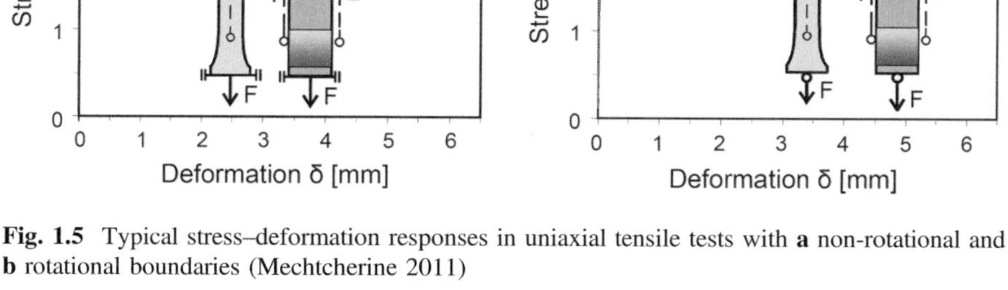

Fig. 1.5 Typical stress–deformation responses in uniaxial tensile tests with **a** non-rotational and **b** rotational boundaries (Mechtcherine 2011)

Comparative testing (Van Zijl et al. 2016) was performed by five laboratories that participated in TC 208-HFC SC2. The SHCC mixes given in Table 1.1 were used. Notably, the casting, curing and test procedures were followed as used in the respective laboratories at the time, and are meticulously described by Van Zijl et al. (2016). All the laboratories concerned used dumbbell-shaped specimens, although of different geometries and size, varying from a gauge area of cross-section 16 × 30 mm, to one that was 30 × 60 mm, and another that was 40 × 40 mm. The gauge length ranged from 80 to 120 mm. Two groups used rotationally fixed boundaries, with one using rotationally free boundaries, and with the other being fixed at one end and free at the other.

Average and maximum crack widths, as well as average crack spacing, which were observed by the five laboratories for the respective laboratories, are summarised in Fig. 1.6 as a function of average tensile strain. Crack width measurement is described in Sect. 1.4.2, although a precursor of such is that the coarser crack measurement resolution of 50 µm in one series by Laboratory 1 (Leipzig FS1), compared with about 10 µm in all other cases, appears to have led to significantly larger crack width observation data, as can be seen in Fig. 1.6a. While detailed results can be found in Van Zijl et al. (2016), an observation of note is that the coefficient of variation (CoV) in crack width was significantly smaller for the larger specimen sizes of L1 and L3, at 0.24–0.59, versus up to 0.90 for the smaller specimen sizes. Small size specimens of L5 were the exception. Various factors could have contributed to the large CoV, including variable fibre dispersion, along with test procedure and set-up, including the prevailing specimen boundary conditions. However, smaller sections might not have been representative elementary volumes in terms of flaw distribution, leading to the larger scatter. Relatively large specimens are preferable and more suitable for subsequent durability testing. Thin

Table 1.1 SHCC mixture proportions used by participating laboratories

Participant laboratory	Cement		Fly ash	Water	Sand		PVA fibre	
	Amount (kg/m^3)	Type	(kg/m^3)	(kg/m^3)	(kg/m^3)	Max. particle size (mm)	Amount (kg/m^3)	Length (mm)
Fine sand (FS)								
L1. Leipzig	550	CEM I 42.5	650	373	550	0.25	26	8
L2. Qingdao	550	CEM I 42.5	650	395	550	0.30	26	12
L3. Rio de Janeiro	550	CEM II F-32	650	395	550	0.20	26	12
L4. Stellenbosch	550	CEM I 42.5	650	395	550	0.30	26	12
L5. Sendai	550	CEM I 52.5	650	395	550	0.20	26	12
Coarse sand (CS)								
L1. Leipzig	560	CEM I 42.5	690	417.6	460	1.00	26	8
L3. Rio de Janeiro	560	CEM II F-32	690	440	460	1.18	26	12
L5. Sendai	560	CEM I 52.5	690	440	460	1.20	26	12

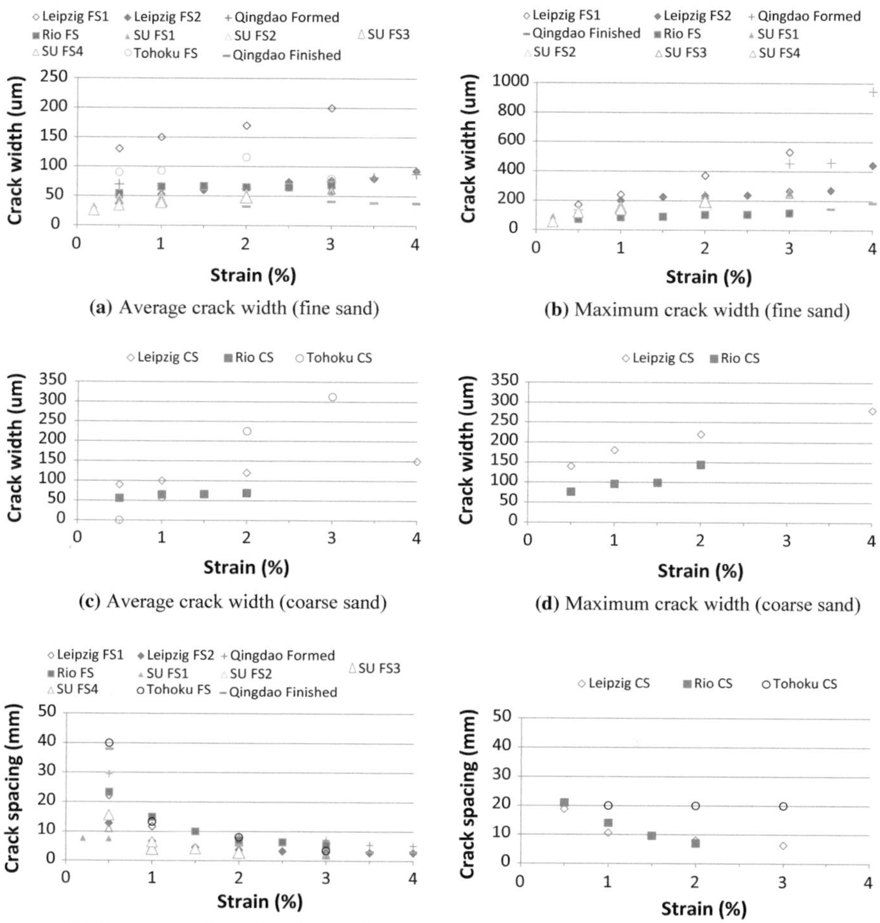

(a) Average crack width (fine sand)
(b) Maximum crack width (fine sand)
(c) Average crack width (coarse sand)
(d) Maximum crack width (coarse sand)
(e) Average crack spacing (fine sand)
(f) Average spacing (coarse sand)

Fig. 1.6 **a** Average and **b** maximum crack widths in SHCC specimens containing fine sand, **c** average and **d** maximum crack widths in coarse sand SHCC, and average crack spacing in **e** fine sand and **f** coarse sand SHCC, as tested by five different laboratories (Van Zijl et al. 2016)

specimens are appropriate for thin-layer applications, provided that sufficient care is taken that the specimens involved are well processed and representative.

The larger specimens had rotationally fixed boundaries at each end, whereas rotationally semi-free ends were used with the smaller specimens, thus preventing a clear distinction to be made between the influence of boundary conditions and that of specimen size in the test results. Carefully applied fixed–fixed boundaries should be preferred in order to activate most of the material's strain capacity. However, if it is impossible to accommodate geometrical imperfections in specimens and adapters, a set of rotationally fixed-free boundaries is recommended, being fixed at one end, and free to rotate at the other.

In the test series, cracks were measured on the *formed surface*, which is the smooth surface facing the mould base during casting. In one laboratory (L2, Qingdao) the crack widths were also measured on the *finished surface*, being the upper surface during casting. Both the average and the maximum crack widths were found to differ significantly on the two surfaces in question, thus requiring further investigation.

Average crack spacing is shown in Fig. 1.6e, f. The average crack spacing reduced from roughly 7 to 40 mm at 0.5% strain, to a stable spacing of 2–7 mm at and beyond 3% strain. The occurrence of crack saturation at roughly 3% strain held true for both fine sand and coarse sand SHCC, although a lower ductility was apparent for the latter. The coarse sand specimens from Sendai (L5) were the exception, maintaining an average spacing of 20 mm, because coarse aggregates (1.2 mm) acted as material defects in the comparatively thin specimens so as to cause localisation at low strains.

Based on the reports of Mechtcherine (2007), Mechtcherine and Jun (2007) and Van Zijl et al. (2016), the following recommendations were made for the tensile testing of SHCC towards durability assessment:

- The SHCC tensile specimen size of thickness of 30–40 mm, of width of 60–100 mm, and of length (of the prismatic section) of 120 mm was recommended for pre-cracking or simultaneous durability testing. This recommendation was based on the relatively large scatter of results that was observed in the relatively small sections, as well as on the practical size of the durability test specimens. Although specimen size may moderately influence average mechanical properties, fibre orientation may have exerted a significant influence on the properties concerned.
- The tensile test set-up should consist of two rotationally fixed ends. Greater uniform strain distribution was induced in the cross-section to a large extent, in contrast to that which was induced by the rotational boundaries.
- A quasi-static loading strain rate, in the range of 10^{-5} to 10^{-3} s^{-1}, should be applied. Various authors have reported rate-dependent tensile stress and strain results for SHCC, ranging from no or insignificant to moderate rate-dependent increased tensile strength in this range of strain rates (Adendorff et al. 2010; Mechtcherine 2011). In the dynamic range, at and beyond 10^{1} s^{-1}, a more significant rate-dependent tensile strength increase was reported by Mechtcherine (2011).

1.4.2 Crack Measurement

In order to characterise durability performance, crack pattern characterisation was required at various levels of deformation, for which ingress and deterioration processes require study. The multiple cracking nature further contributed to significant data gathering. Contactless methods were advised, with photogrammetric analysis to this purpose having already been described in recent relevant literature (Boshoff and Adendorff 2013; Wagner et al. 2012). Figure 1.7 shows the crack observation equipment that was used by each group participating in the comparative testing.

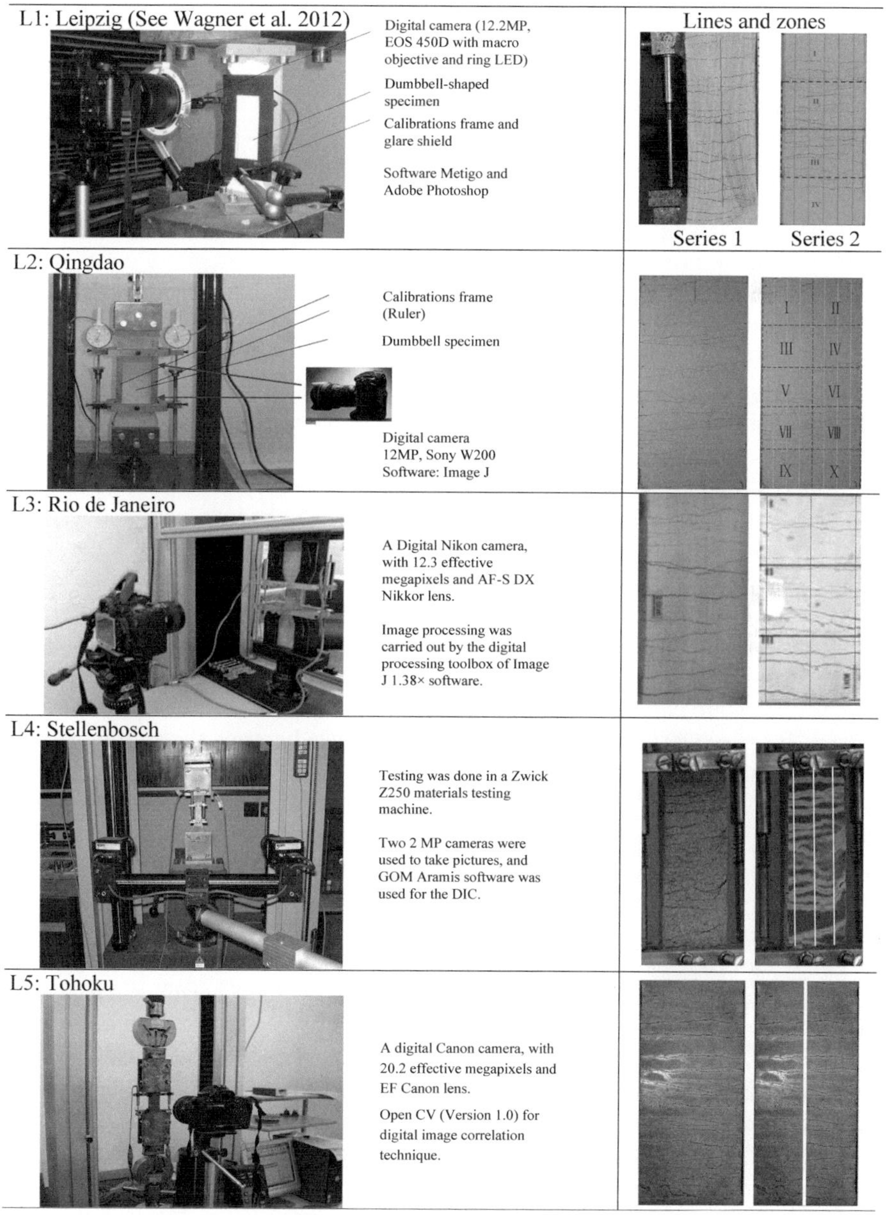

Fig. 1.7 Crack observation and measurement methods (Van Zijl et al. 2016)

Contactless photographic crack characteristic measurement procedures can be classified into two categories, namely digital image processing (DIP) and digital image correlation (DIC). Although DIP is typically more direct, such processing requires additional human input and further assumptions than does the more automated DIC. Both methods have been shown to give accurate crack width measurements. Notably, for both methods only the crack characteristics on the surface of one side of the specimens were measured.

Although not a true representation of the full crack pattern within the specimens, due to cracks typically merging and branching, the above does, however, give a good indication of the crack pattern induced. DIP is a direct method for determining the crack widths and positions by means of evaluating the pixels of the photograph taken. The photograph is scaled using a known reference, with it being possible to make a correction to compensate for the distortion (aberration) found relatively close to the edges of the photo, due to lens effects. The photograph concerned is typically converted to grey scale, with a certain level of white intensity being chosen to indicate the edge of a crack in the direction perpendicular to the crack. The number of pixels between the start and the end of a crack is then counted and used to calculate the crack widths involved. The choice of the level of intensity of the pixel that indicates the start of the crack can, however, influence the measured crack widths. Alternatively, the edge of a crack can be manually determined by the user, although doing so is less objective than the former method.

The DIC is an indirect measurement of the crack widths involved. A distinctive speckle/stochastic pattern is required on the surface of the specimen. One, but preferably two, cameras are used to take pictures of the specimen in the reference (i.e. uncracked) condition, and then at the required strain levels. A correlation algorithm is used to analyse the surface after dividing the surface into a regular pattern of facets. The algorithm then determines the original and deformed coordinates (which are two-dimensional for one camera, and three- dimensional for two cameras) on the surface of the specimen. The distance between the coordinates can be prescribed, although a distance of 1 mm is often used. The relative change in distance between coordinates in a direction perpendicular to the crack is used as an indication of whether a crack has formed between two coordinates. It is, therefore, vital to choose the coordinate system with one axis in the direction of the cracks. The increase of distance between the coordinates is then used as the crack width. The advantage of DIC is that it is relatively easy to use, as well as being accurate. However, the cracks are not measured directly, and when two cracks occur between two coordinates, the combined width amounts to the calculated crack width. Also, if the speckle pattern is severely distorted by means of a crack, the correlation algorithm might not be able to determine the deformed coordinates at a specific position, and the commercially available software will then typically interpolate, thus giving false crack width readings.

Different crack measurement methods were followed in the comparative test programme, namely DIP in the case of L1, L3 and L5, using high-resolution cameras and the standard commercial DIP software, as listed in Fig. 1.7, while, in the case of L2 and L4, DIC was used. In Leipzig (L1), the first series of crack measurement (series FS1 and CS1) was done with low resolution, in the order of 50 μm. In series FS2, as undertaken in the laboratory in question, an improved resolution of 10 μm was achieved. In Qingdao (L2), 2D photos were taken with a resolution of approximately 10 μm. In Stellenbosch (L4), 3D Aramis DIC was used, with a 10 μm resolution. Hence, in Laboratories L1 (series FS2), L2, L3, L4 and L5, approximately the same resolution of about 10 μm was achieved.

Multiple cracks were observed by all the laboratories concerned. The cracks were not all parallel or continuous throughout the specimen width. Crack widths were determined at various positions along the specimen length, along a central vertical line by L1, in their first series (FS1 and CS1), and in L5, along three vertical lines by L3 and L4, along four vertical lines by L2, and along five vertical lines in the second series (FS2) by L1. Each crack that crossed either of the lines was captured and its width measured. From the cracks, the spacing, average and maximum width were determined, as is shown in Fig. 1.6, and the frequency distributions derived. The results obtained are presented in the next section. Based on the comparative test programme results the use of contactless methods, preferably DIC with a resolution of, at most, 10 μm, is recommended.

1.4.3 Crack Pattern Representation

A useful presentation of crack width data is shown in Fig. 1.1b, in the form of crack width histograms. Such representation allows for the linking of crack distributions with deterioration resistance. In addition, average crack widths per set, standard deviation, and maximum crack widths, as well as average crack spacing, must be reported, although all the values concerned might be derived from the histogram data with reasonable accuracy. The durability of cracked SHCC, as dependent on the crack pattern, is subsequently reported throughout the chapters of the current STAR, for various deterioration processes, including for the corrosion of steel reinforcing bars embedded in cracked SHCC (in Chap. 9).

An alternative way of presenting crack width distributions is the so-called crack width polygon shown in Fig. 1.8a (Wagner et al. 2012). The format of such curves resembles the one of grain size distribution curves, and it allows presentation of the crack width distribution independent of such further crack pattern characteristics as the crack density (CD), spacing and length. For the two last-mentioned characteristics, the distributions of the values can be represented in the same way. It is also possible to characterise such distributions by means of single numerical values based on the polygons involved (see Chap. 2). In the case of the crack width distribution, the single value is called the *crack width value* (*CWV*). The reason for a separate consideration of crack pattern characteristics is located in their different modes of action with regard to transport, or degradation, processes. Depending on the respective load, the single values can be weighted with physical background data, and hereby linked to the observed process. An example is shown in relation to the water permeability of cracked SHCC in Fig. 1.8b. Using the weighted *CWV*

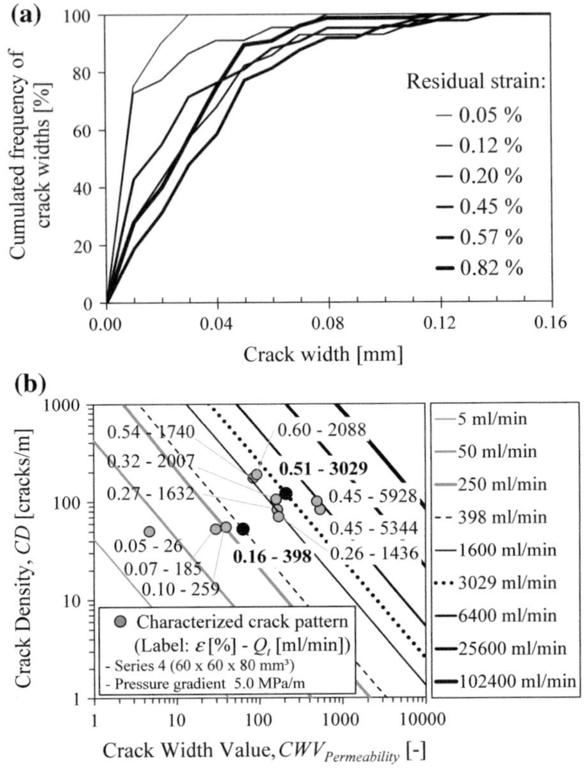

Fig. 1.8 **a** Crack width polygons and **b** crack density and cumulated crack width representations, with isolines for the theoretical water flow rate, as proposed by Wagner et al. (2012)

($CWV_{\text{Permeability}}$) and the CD, the theoretical water permeability of the cracks can be determined for a given pressure gradient. The crack spacing distribution appears to have a significant influence on the capillary absorption of cracked SHCC (see Chap. 2). As elaborated on in Chap. 9, crack spacing is also of significance in the deterioration process of corrosion.

1.5 Applications of SHCC Towards Enhancing Durability

Various structural applications of SHCC have arisen in this century, as reported by various authors. SHCC characteristics of

i. mechanical damage resistance, as well as freeze–thaw cycle deterioration resistance are demonstrated in its superior behaviour in *patch repair* application, in comparison to that in traditional patch repair materials (Wagner and Slowik 2012);
ii. significant *tensile and flexural ductility* are key to the integrity and durability of recent highway bridge connecting slabs (Muroda et al. 2013; Ooyagi et al. 2013),

highway bridge joint systems (Ishikawa et al. 2014), and the so-called link slab replacing a bridge movement joint in Michigan (Lepech and Li 2005, 2009b);

iii. *toughness and energy dissipation* through multiple crack formation have been realised in the coupling beams of high-rise building cores in seismic regions (Rokugo 2005; Rokugo and Kanda 2013), and

iv. *compatible tensile deformation with embedded reinforcing steel*, as illustrated in Fig. 1.2, are essential in crack control and retained steel–SHCC interfacial bond in such R/SHCC applications as the link slab (Lepech 2011; Lepech and Li 2005), the deformable connecting slab (Muroda et al. 2013; Ooyagi et al. 2013), and the highway bridge RC plug joint (Ishikawa et al. 2014).

Types of applications i, ii and iv are elaborated on in the following sections, for their particular relevance to structural durability.

1.5.1 Patch Repair, Leipzig, Germany

Wagner and Slowik (2012) report on patch repair with SHCC to a depth of 40 mm around inspection holes at a fuel station belonging to a construction company in Saxony, Germany, as well as in a separate 5 × 12 m area, as is shown in Fig. 1.9. The area carries heavy vehicles (trucks, tracked vehicles, forklifts, etc.), with the areas shown being heavily damaged. Shafts 1 and 2, as well as the larger of the rectangular areas, were repaired with SHCC, whereas a polymer-modified cement mortar was used in Shaft 3.

Five years after the repair measure, only Shaft 2 and the larger patch remained, whereas Shafts 1 and 3 had been completely replaced. Figure 1.10 compares the appearance directly after repair and five years later. After the stated period, adhesive bond strength tests were performed on the repair patches, with the interface strength

Fig. 1.9 Patch repair preparations around **a** inspection shafts and **b** a 2 × 5 m patch (Wagner and Slowik 2012)

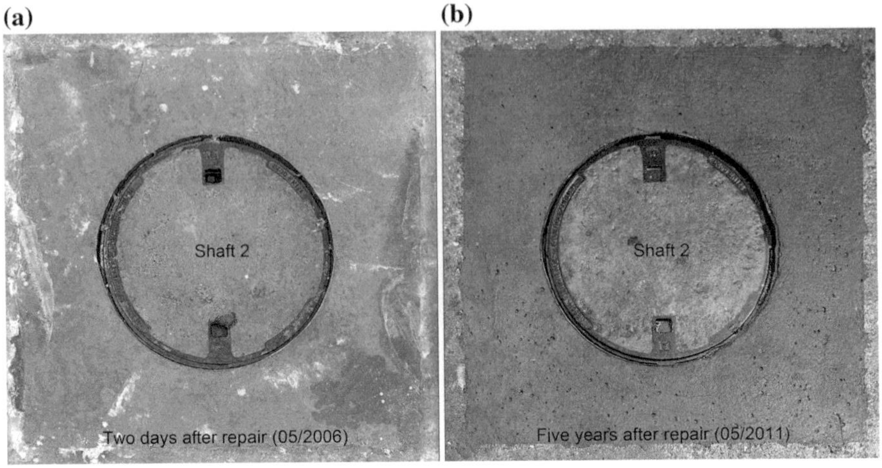

Fig. 1.10 40-mm-thick SHCC patch repair condition **a** directly after repair and **b** 5 years later (Wagner and Slowik 2012)

between the SHCC layer and the concrete substrate being determined to be 1.3 N/mm^2. However, in most pull-off tests, the concrete failed, despite the moderate strength of the materials involved. Cores were drilled from the SHCC repair layer. A dry density of 1570 kg/m^3 and an average compressive strength of 74.5 N/mm^2 could be measured (core diameter 50 mm, height approximately 40 mm). Note that the low height to diameter aspect contributed to the high apparent compressive strength. The carbonation depth was less than 1 mm, although it was slightly larger at the edges of the cracks. Surface crack pattern characterisation was made difficult due to the blunt crack edges and the dust that were present. An average crack spacing of about 35 mm between tangential cracks and 27 mm between radial cracks was observed.

The pilot case study, which lasted over a period of five years, proves that SHCC can be successfully used for patch repair. Not only was the patch in question intact, but the crack widths appeared to be limited, and the evidence of carbonation degradation was restricted to a fraction of the patch depth.

1.5.2 *Repair of a Water Reservoir of the Hydraulic Power Pumping Station in Thüringen, Germany*

A number of very promising applications of SHCC in terms of repair, as well as in the terms of protection of hydraulic structures, have already been demonstrated. In 2003, the Mitaka Dam, a gravity concrete dam in Hiroshima, Japan, was repaired (Kojima et al. 2004). Thirty cubic meters of SHCC was sprayed onto an area of 500 m^2 on the upstream side of the dam, to a thickness of 30 mm, so as to improve the waterproofing performance of the deteriorated concrete surface. An anchor was driven for each 1.5 m^2 to ensure the stability of the bond between the base concrete and the SHCC.

A similar measure was tested in Thüringen, Germany in 2011, when SHCC, which had been developed at the TU Dresden, was used to repair the upper water reservoir of the hydraulic power pumping station, Hohenwarte II; however, in this case no anchorage system was used. Before repair, the reservoir wall showed cracks, defective joints, scaling of the concrete, and other defects. The purpose of the pilot project was to restore the water tightness of the wall in just one working step, using a single, sprayed layer of SHCC (see Fig. 1.11). Due to its ductility, the repair material used should have durably sealed the defective joints and cracks, in addition to which the SHCC should have provided a smooth, durable cover for the concrete substrate. Since the reservoir walls were exposed to high water saturation and to considerable changes in water level, the danger of frost damage was considerable. Although the use of conventional air-entrainment agents should have been possible with SHCC, it was not so when the spraying technique was applied. For this reason, superabsorbent polymers (SAP) were introduced as a new additive to improve the frost resistance of the repair layer. More information on the use of SAP in concrete construction can be found in Mechtcherine and Reinhardt (2012).

In the summer (for the northern hemisphere) of 2015, being four years after the initiation of the operation, the condition of the repair layer was comprehensively inspected, including taking cores. Whereas the horizontal joints had had openings of several millimetres before the repair, multiple fine cracks with widths ranging between 0.05 and 0.1 mm had developed, as had been expected, in the SHCC layer over the joints in question. Furthermore, it could be observed that the water content and the soiling of the joint regions had been significantly reduced by means of the repair measure taken.

(a) (b)

Fig. 1.11 **a** Application of SHCC containing SAP as repair material for spraying (×3), **b** condition of the surface after completion of the repair work (courtesy of Professor Viktor Mechtcherine, TU Dresden)

1.5.3 Deformable Connecting Slab, Tanabe, Japan

A new slab connection system consisting of R/SHCC, as illustrated in Fig. 1.12a, was developed, tested and proposed for implementation in the Tanabe parking area (Muroda et al. 2013; Ooyagi et al. 2013). This was done to: (i) produce a continuous system avoiding maintenance-prone movement joints, facilitating construction, and (ii) avoid transfer of significant forces, such as is experienced during seismic events in the case of the conventional connection type that is shown in Fig. 1.12b. The material used was designed for a compressive strength of 30 MPa, with relatively low elastic modulus, in the range of 15–20 GPa, high ductility, and dispersed cracking.

In Fig. 1.12c, the deflection-hardening behaviour of the connecting slab material is shown, with typical indications of multiple crack formation in the form of wiggles in the load-deflection curve. On the right, in Fig. 1.12d, the construction of such a connecting slab is shown, with the reinforcing steel, the standard construction equipment and the flowable SHCC used being clearly visible.

The above-mentioned concept has been installed and operational since 2011. The continuous nature of the material used allowed for the waterproofing and asphalt to be placed over the connecting slab, which meant that the condition of the R/SHCC connecting slab itself could not be monitored. However, when it was last

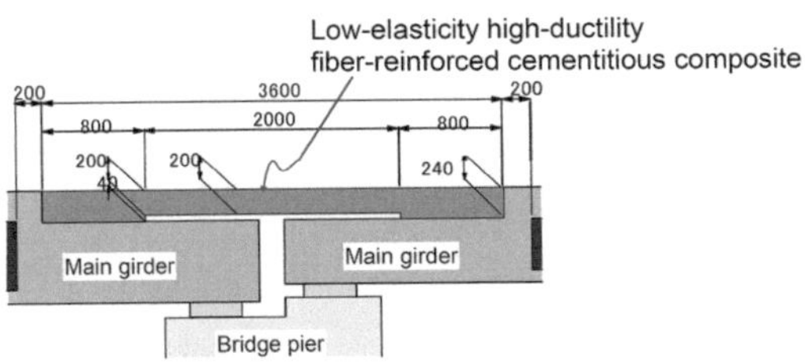

(**a**) Muroda et al. (2013)

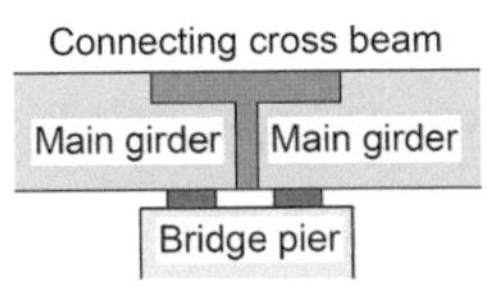

(**b**) Muroda et al. (2013)

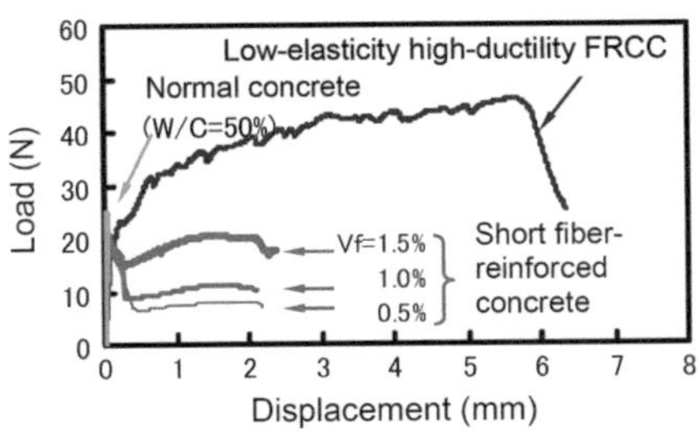

(**c**) Ooyagi et al. (2013)

(**d**) Taniguchi et al. (2011)

Fig. 1.12 Slab connection system schematic, showing **a** the new connection type (Muroda et al. 2013), **b** the traditional connection (Muroda et al. 2013), **c** the deflection-hardening plate flexural response (Ooyagi et al. 2013), and **d** the casting of a connecting slab on-site (Taniguchi et al. 2011) examined, it appeared to be in good condition, due to the lack of visible cracks and potholes in the asphalt.

1.5.4 Jointless Bridge Slab–Abutment Link, Nippon Highway, Japan

In a step taken towards improving the durability of infrastructure against the de-icing salt-induced corrosion, Ishikawa et al. (2014) report the development of a jointless bridge abutment–slab-connected system enabled by the use of highly ductile cement-based material. The system has been termed the RC plug joint by the developing authors. The use of traditional joints between abutments and suspended slabs on bridges has led to a significant amount of leakage and corrosion deterioration in the RC abutments and slabs beneath the traditional joints in highway bridges. By replacing such leaking joints with steel-bar-reinforced highly ductile fibre-reinforced composites, capacity for thermal expansion and shrinkage is retained. Sealing is also thereby improved, because the jointless connection allows for continuous deck waterproofing and asphalt layering over the region in question.

The concept is schematised in Fig. 1.13a, with the casting process being shown in Fig. 1.13b. Ishikawa et al. (2014) report that 85 joint replacements of the above-mentioned type have already been performed in Japan. In monitoring performance, delamination had been found to cause leakage in a limited number of cases (3). In addition to the significantly reduced amount of leakage that had taken place, velocity response to passing test vehicle was recorded as being significantly reduced after the installation of the new jointless link, in comparison with the traditional joints. Such a response may be interpreted as enabling improved driving comfort, as well as potentially improved durability and fatigue life of the bridges concerned.

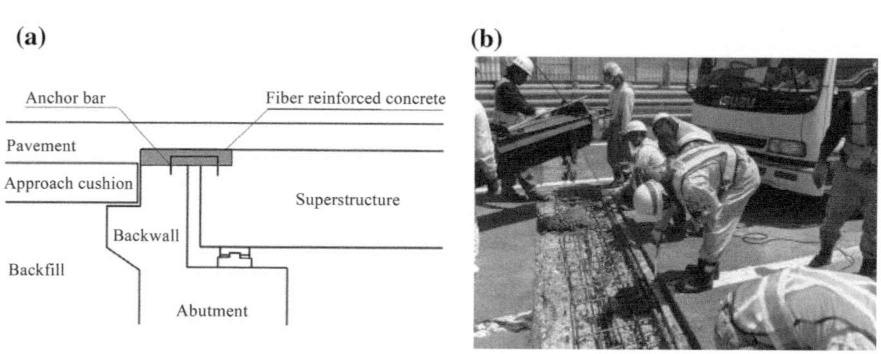

Fig. 1.13 Jointless abutment–slab connection towards the prevention of leakage and corrosion deterioration in highway infrastructure, showing **a** the schematised connection and **b** the casting of the replacement seamless joint (Ishikawa et al. 2014)

1.5.5 Link Slab, Michigan, USA

In 2005, in Michigan, USA, expansion joints of a highway bridge were replaced by means of an R/SHCC link slab (Lepech and Li 2005, 2009b). The R/SHCC link slab elements were designed to eliminate problematic expansion joints on multi-span vehicle bridges. As Michigan experiences harsh winters, de-icing salt practice is followed on the highways. In the present instance, leakage of the traditional bridge deck expansion joint caused the supporting beam ends to corrode, leading to the need to replace the non-functioning traditional joint as frequently as every three years. The replacing R/SHCC link slab system, which is illustrated in Fig. 1.14, is expected to have a significantly longer service life than was the case with the previously used material (Lepech and Li 2005).

The retrofitting procedure was performed in 2005, and, after a decade of continued de-icing practice, the link slab in question is currently still operational. However, cracks have been observed on the surface of the link slab, indicating its intended extension action by means of multiple crack formation.

In this link slab application, the abrasion resistance of SHCC appears to have been sufficient, which is a topic that is described in more detail in Chap. 7.

1.6 Modelling

To facilitate prediction and durability design, modelling of both crack widths and crack spacing is required. For research purposes towards linking crack patterns and deterioration, it has been recommended that crack width distributions be presented

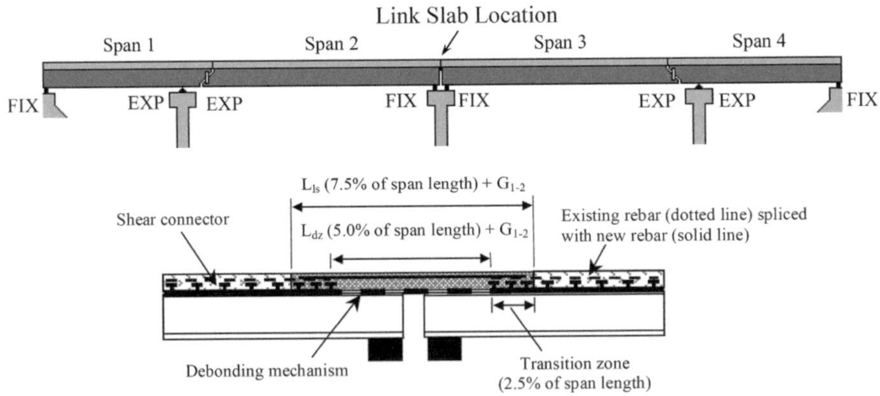

Fig. 1.14 Link slab schematisation towards replacement of problematic movement joints in multi-span highway bridges subject to de-icing salt deterioration (Lepech and Li 2005)

as shown in Fig. 1.1b, with crack width polygons shown in Fig. 1.8a, or that descriptive statistical distributions be used, such as those that are described by Boshoff et al. (2016). Crack spacing should also be presented accordingly, to assist, for instance, in reinforcing bar corrosion evaluation, as discussed in Chap. 9.

Cracking data were given in Fig. 1.6 of this chapter as made available from recent comparative testing (Van Zijl et al. 2016). In previous works, summaries of crack width and spacing have been given for other SHCCs from the literature (Van Zijl and Wittmann 2011; Van Zijl et al. 2012). While such widths and spacings might serve as a guideline, it must be kept in mind that they are likely to vary depending on the particular SHCC mix used in the project for which the durability design is to be performed.

The two approaches to modelling crack widths and patterns consist of analytical modelling, such as micromechanical modelling, and empirical modelling, such as representing the crack widths using descriptive statistics.

From micromechanical modelling and experimental characterisation (Li 1993; Li et al. 1990), the following expression for crack opening in SHCC at maximum fibre resistance was proposed:

$$w_c = \frac{\tau L_f^2}{E_f d_f \left(1 + V_f E_f / V_m E_m\right)} \tag{1.1}$$

with:

- τ the interfacial bond strength between the fibre and the matrix
- L_f the fibre length
- d_f the fibre diameter
- V_f the fibre volume fraction
- V_m the matrix volume fraction $= 1 - V_f$
- E_f the fibre elasticity modulus
- E_m the matrix elasticity modulus

Notably, this expression does not predict the crack width at typical serviceability strain levels (e.g. at less than 1%), but at the maximum fibre resistance. For typical polyvinyl alcohol (PVA) fibre SHCC values (τ = 1 MPa, L_f = 12 mm, d_f = 0.04 mm, V_f = 0.2, E_f = 40GPa, E_m = 15 GPa), the calculated crack width is around 85 μm. The model also does not include the effect of slip-hardening that is typically found for PVA fibres (Lin and Li 1997).

The above-mentioned model has been further developed by Cai et al. (2011) as including the statistical distribution of the fibres, and as also including the flaw size and its distribution. The model was found to correlate well with experimental results. Cai et al. (2011) also developed a simplified expression for the typical spacing between cracks in SHCC (L_{eff}), namely:

$$S_c = L_f\left(1 - \sqrt{1 - \frac{2\sigma_{mc}d_f}{gV_f\tau_0 L_f}}\right) \tag{1.2}$$

with:

σ_{mc} the matrix cracking strength
τ_0 the frictional sliding shear stress at the tip of the debonding zone where no slip occurs
L_f the fibre length
d_f the fibre diameter and
V_f as previously defined

The above formulation does not, however, include the statistical distribution of the initial flaw size and the chemical bond between the fibre and the matrix. Numerous studies, including those of Boshoff et al. (2016) and Wang et al. (2011), have been done on the statistical description of the crack widths of SHCC. Such expressions can be useful for the durability design of SHCC, as noted by Boshoff et al. (2016), among others. A log-normal distribution with a zero lower boundary has been proposed by Wang et al. (2011) and Boshoff et al. (2016), whereas Nieuwoudt (2012) found that the best distribution for the specific SHCC evaluated was that of a Gamma distribution, with a 20 μm lower boundary. It is, however, clearly reported by all authors concerned that a normal distribution is not appropriate for modelling the crack width distribution.

The empirical modelling of the crack spacing, which is also important, should take into account not only the average distance between cracks, but also the local grouping of cracks. Boshoff and Nieuwoudt (2011) have attempted to model the grouping of cracks empirically, using a so-called Crack Proximity Index (CPI). An example of different crack patterns on tensile specimens for the same strain level (1%) is shown in Fig. 1.15, together with the calculated CPI for each sample at different strain levels.

In RC, crack spacing is generally considered to be larger for larger cover to steel (EN 1992-1-1:2004), as considered in a recent reliability study by McLeod et al. (2013). In Chap. 9, in the discussion on steel corrosion in cracked R/SHCC, an opposite trend is reported. The inherent fine crack spacing in SHCC, as predicted by Eq. 1.2, and as seen in SHCC that is subjected to tension or flexure, appears to be influenced in R/SHCC by the ribs of steel reinforcing bars. Such can be seen in Fig. 1.2b, which shows crack spacing at the steel bar interface being dominated by the deformation pitch, or spacing. Farther away from the bar, cracks split and result in finer spacing at a distance from the bar. Paul and Van Zijl (2013) have reported a larger crack spacing in R/SHCC beams for cover depth of 15 mm, than for cover depths of 25–35 mm. Nevertheless, it must be emphasised that the fine crack spacing that characterises SHCC is retained in R/SHCC, as is clearly evident in Fig. 1.2.

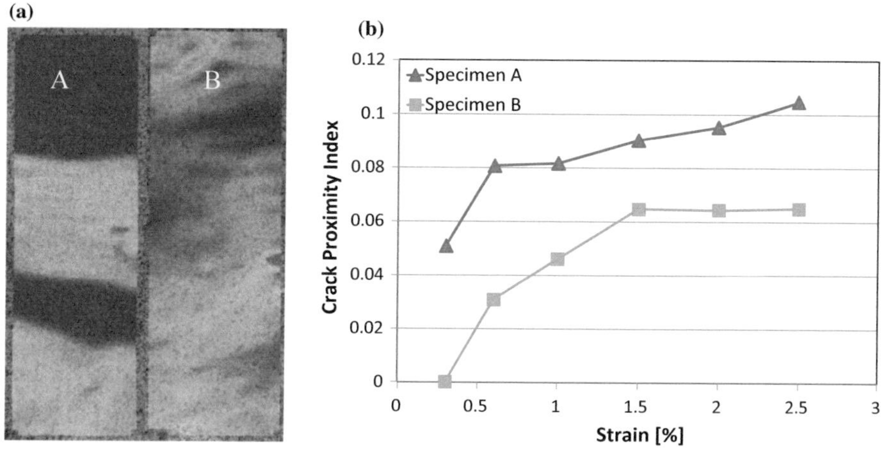

Fig. 1.15 **a** Crack patterns of two tensile SHCC specimens, both at 1% strain, **b** the calculated Crack Proximity Index for the same two specimens at different strain levels (Boshoff and Nieuwoudt 2011)

1.7 Conclusions

This chapter has introduced the notion of crack formation and durability of SHCC, which is known for forming multiple cracks under imposed strain. Following the STAR published by RILEM TC 208-HFC SC2 (Van Zijl and Wittmann 2011), the new TC 240-FDS has, since 2010, continued the work towards characterising crack patterns under imposed strain and durability performance in the operating (i.e. cracked) state when exposed to various deteriorating conditions.

Particular matters identified as research questions by the previous TC 208-HFC SC2 were stated and briefly described in this chapter. The occurrence of wider cracks in SHCC, despite fine average crack widths, was identified as a particular matter for investigation. Such investigation should be aimed at understanding what the durability implications are of the presence of individual wider cracks, which could reach widths similar in magnitude to typical threshold crack widths in reinforced concrete. A second matter of research interest lies in the high rate of ingress into finely cracked SHCC by capillary absorption. Research results on these matters are elaborated on in subsequent chapters of this STAR.

Recommended practice in executing direct tension pre-cracking of SHCC specimens intended for durability tests, or during which simultaneous exposure to deterioration agents is to be performed, has been summarised in this chapter. Specimen shape and size, fixed boundary conditions in the materials testing machine, and test rate have been recommended, based on prior work by TC 208-HFC SC1 (Mechtcherine 2011) and a comparative test programme by members of TC 208-HFC SC2. Four of the five laboratories participating in TC 208-HFC

SC2 have joined the new RC 240-FDS, furthering this study to the point of final recommendation and publication (Van Zijl et al. 2016).

Finally, the application of SHCC in actual structures, in patch repair, and in the structural applications of both connecting slabs and abutment-slab linking elements in highway bridge infrastructure exposed to heavy winter and salt de-icing practices have been reported. Although several other applications have already been reported in the STAR TC 208-HFC SC3 (Rokugo and Kanda 2013), and a particular link-slab application reported by SC2 of that TC (Lepech 2011), the above three applications were selected to highlight the durability enhancement of infrastructure by means of the sound use of SHCC.

References

Adendorff, C.J., Boshoff, W.P., Van Zijl, G.P.A.G. (2010). Crack characterisation in SHCC: towards durability assessment. Proceedings of the International Conference on Advanced Concrete Materials, 17-19 November 2009, Stellenbosch, South Africa, CRC Press, London, pp. 215-221.

Boshoff, W.P., Adendorff, C.J. (2013). Effect of sustained loading on SHCC crack widths. Cement and Concrete Composites 37:119-125.

Boshoff, W.P., Altmann, F., Adendorff, C.J., Mechtcherine, V. (2016). A new approach for modelling the ingress of deleterious materials in cracked strain hardening cement-based composites. Materials and Structures 49:2285-2295.

Boshoff, W.P., Nieuwoudt, P.D. (2011). Tensile crack widths of strain hardening cement-based composites. Proceedings of the 2nd RILEM Conference on SHCC (SHCC2), 12-14 December 2011, Rio de Janeiro, Brazil, RILEM Publication S.A.R.L., Bagneux, France, pp. 199-206.

Cai, X.R., Xu, S.L., Fu, B.Q. (2011). A statistical micromechanical model of multiple cracking for ultra-high toughness cementitious composites. Engineering Fracture Mechanics 78:1091-1100.

Dvorkin, D., Peled, A., Weiss, W.J. (2013). Influence of bundle coating on the tensile behavior, bonding, cracking and fluid transport of fabric cement-based composites. Cement and Concrete Composites 42:9-19.

EN 1992-1-1 (2004). Eurocode 2: Design of concrete structures – Part 1: General requirements. British Standards Institute.

Fischer, G., Li, V.C. (2002). Reinforced ECC – from materials to structures. Proceedings of the 1st FIB Congress, October 2002, Osaka, Japan, pp. 105-122.

Ishikawa, Y., Aoyama, M., Kuroyanagi, M., Nagai, M., Miyashita, T. (2014). Proposition of a new type of jointless system for existing concrete bridges. Journal of Physical Science and Application 4(2):84-89.

Kabele, P., Takeuchi, S., Inaba, K., Horii, H. (1999). Performance of engineered cementitious composites in repair and retrofit. Proceedings of the 3rd RILEM Workshop on High Performance Fibre-reinforced Cement-based Composites (HPFRCC3), 16-19 May, Mainz, Germany, RILEM Publications S.A.R.L., Mainz, Germany, pp. 617-627.

Kojima, S.N., Sakata, N., Kanda, T., Hiraishi, M. (2004). Application of direct sprayed ECC for retro-fitting dam structure surface-application for Mitaka-Dam. JCI Concrete Journal 42(5).

Lepech, M. (2011). Durability, economical, ecological, and social aspects: life-cycle considerations, Chapter 8 in Durability of strain-hardening fiber-reinforced cement-based composites (SHCC), State-of-the-Art Report of RILEM Technical Committee 208-HFC, Sub-Committee 2.

Lepech, M., Li, V.C. (2005). Design and field demonstration of ECC link slabs for jointless bridge decks. Proceedings of the 3rd International Conference on Construction Materials:

Performance, Innovations, and Structural Implications (ConMat'05), 22-24 August, Vancouver, Canada.
Lepech, M., Li, V.C. (2009a). Water permeability of engineered cementitious composites. Cement and Concrete Composites 31:744-753.
Lepech, M., Li, V.C. (2009b). Application of ECC for bridge deck link slabs. Materials and Structures 42:1185-1195.
Li, V.C., Wang, Y., Backer, S. (1990). Effect of inclined angle bundling and surface treatment on synthetic fiber pull-out from a cement matrix. Composites 21(2):132-140.
Li, V.C. (1993). From micromechanics to structural engineering – the design of cementitious composites for civil engineering applications. Journal of Structural Mechanics and Earthquake Engineering 10:37-48.
Lim, Y.M., (1996). Interface fracture behavior of rehabilitated concrete infrastructures using engineered cementitious composites. Dissertation, Michigan University, USA.
Lin, Z., Li, V.C. (1997). Crack bridging in fiber reinforced cementitious composites with slip-hardening interfaces. Journal of the Mechanics and Physics of Solids 45(5):763-787.
McLeod, C., Retief, J.V., Wium, J.A. (2013). Reliability of EN 1992 crack model applied to South African water retaining structures. Proceedings of the International Conference in Structural engineering, Mechanics and Computation (SEMC 2013), 2-4 September, Cape Town, South Africa, Taylor and Francis, London, pp. 1549-1554.
Mechtcherine, V. (2007). Testing behaviour of strain hardening cement-based composites in tension – summary of recent research. Proceedings of the 5th RILEM Symposium on High-Performance Fibre Reinforced Cementitious Composites (HPFRCC5), 10-13 July, Mainz, Germany, RILEM Publications S.A.R.L., pp. 13-22.
Mechtcherine, V. (2011). Tensile tests, Chapter 3 in Mechanical characterization and testing of fiber-reinforced cement-based composites (SHCC). State-of-the-Art Report of RILEM Technical Committee 208-HFC, Sub-Committee 1.
Mechtcherine, V., Jun, P. (2007). Stress-strain behaviour of strain-hardening cement-based composites (SHCC) under repeated tensile loading. Proceedings of the International Conference on Fracture Mechanics of Concrete and Concrete Structures (FRAMCOS-6), 17-22 June 2007, Catania, Italy.
Mechtcherine, V., Jun, P. (2008). Behaviour of strain hardening cement-based composites in tension and compression – overview of recent research. Proceedings of the 7th International RILEM Symposium on Fibre Reinforced Concrete: Design and Applications, RILEM Publications S.A.R.L., pp. 471-481.
Mechtcherine, V., Reinhardt, H.W. (eds.) (2012). Application of superabsorbent polymers in concrete construction. RILEM State of the Art Reports, Springer.
Mechtcherine, V., Lieboldt, M. (2011). Permeation of water and gases through cracked textile reinforced concrete. Cement and Concrete Composites 33:725-734.
Muroda, K., Taira, Y., Nabetani, Y., Iijima, M., Uchida, S., Oshiro, T. (2013). Development and practical application of slab connection system for PC girders using low-elasticity high ductility fiber-reinforced cementitious composite. Proceeding of the 3rd Sustainable Construction Materials and Technologies Conference (SCMT3), 18-21 August 2013, Kyoto, Japan, Japan Concrete Institute.
Nieuwoudt, P.D. (2012). Quantifying the cracking behaviour of strain-hardening cement-based composites. MSc Thesis, Stellenbosch University, South Africa.
Ooyagi, R., Taniguachi, H., Sasaki, W., Higuchi, M., Shiina, N., Oshiro, T. (2013). Development of low-elasticity high-ductility fiber-reinforced cementitious composite for use in PC girder connecting slabs. Proceedings of the 3rd Sustainable Construction Materials and Technologies Conference (SCMT3), 19-21 August 2013, Kyoto, Japan, JCI.
Paul, S.C., Van Zijl, G.P.A.G. (2013). Mechanically induced cracking behaviour in fine and coarse sand strain hardening cement based composites (SHCC) at different load levels. Journal of Advanced Concrete Technology 11:301-311.
Rokugo, K. (2005). Applications of strain hardening cementitious composites with multiple cracks in Japan. Proceedings of the Symposium on Ultra-ductile Concrete with Short Fibres

Development, Testing, Applications, 25 October 2005, Kaiserslautern, Germany, Ibidem-Verlag, Stuttgart, pp. 121-133.
Rokugo, K., Kanda, T. (eds.) (2013). Strain hardening cement composites: structural design and performance. State-of-the-Art Report of RILEM Technical Committee 208-HFC, Sub-Committee 3, Volume 6, Springer.
Silva, F. de Andrade, Butler, M. Hempel, S. Toleo Filho, R.D., Mechtcherine, V. (2014). Effects of elevated temperatures on the interface properties of carbon textile-reinforced concrete. Cement and Concrete Composites 48:26-34.
Taniguchi, H., Taira, Y., Muroda, K., Oshiro, T. (2011). Construction of the link slab for PC girders using fiber reinforced cementitious composite (in Japanese). Concrete Journal 49 (4):22-29.
Van Zijl, G.P.A.G., Wittmann, F.H. (2010). On durability of SHCC. Journal of Advanced Concrete Technology 8(3):261-271.
Van Zijl, G.P.A.G., Wittmann, F.H. (eds.) (2011). Durability of strain-hardening fibre-reinforced cement-based composites (SHCC), State-of-the-Art Report of RILEM Technical Committee 208-HFC, Sub-Committee 2, Volume 4, Springer.
Van Zijl, G.P.A.G., Wittmann, F.H., Oh, B.H., Kabele, P., Toledo Filho, R.D., Fairbairn, E.M.R., Slowik, V., Ogawa, A., Hoshiro, H., Mechtcherine, V., Altmann, F., Lepech, M.D. (2012). Durability of strain-hardening cement-based composites (SHCC). International Journal of Materials and Structures 45(10):1447-1463.
Van Zijl, G.P.A.G., Wittmann, F.H., Toledo Filho, R.D., Slowik, V., Mihashi, H. (2016). Comparative testing of crack formation in SHCC. International Journal of Materials and Structures 49(4):1175-1189.
Wagner, C., Slowik, V. (2012). Strain hardening cement-based composites for repair layers on cracked concrete surfaces. Proceedings of Concrete Solutions, 4th International Conference on Concrete Repair, 26-28 September 2011, Dresden, Germany, Taylor and Francis Group, London, pp. 775-782.
Wagner, C., Dolase, A., Slowik, V. (2012). Evaluation of crack patterns in SHCC with respect to water permeability and capillary suction. Proceedings of the 3rd International Conference on Concrete Repair, Rehabilitation and Retrofitting (ICCRRR 2012), 3-5 September 2012, Cape Town, South Africa, CRC Press, Leiden, The Netherlands, pp. 972–977.
Wang, K., Jansen, D.C., Shah S.P. (1997). Permeability study of cracked concrete. Cement and Concrete Research 27(3):381-393.
Wang, P., Wittmann, F.H., Zhao, T., Huang, W. (2011). Evolution of crack patterns on SHCC as function of imposed strain. Proceedings of the 2nd RILEM Conference on SHCC (PRO 81), 12-14 December 2011, Rio de Janeiro, Brazil, RILEM Publication S.A.R.L., Bagneux, France, pp. 217-224.
Wittmann, F.H., Zhao, T., Wang, F., Wang, L. (2010). Aspects of durability of strain hardening cement-based composites under imposed strain. Proceedings of the International Conference on Advanced Concrete Materials, 17-19 November 2009, Stellenbosch, South Africa, CRC Press, pp. 173-179.
Zhang, P., Wittmann, F.H., Zhao, T.J., Lehmann, E.H., Tian, L., Vontobel, P. (2010). Observation and quantification of water penetration into strain-hardening cement-based composites (SHCC) with multiple cracks by means of neutron radiography. Nuclear Instruments and Methods in Physics Research A 620 (2010):414-420.

Chapter 2
Transfer of Fluids, Gases and Ions in and Through Cracked and Uncracked Composites

Christian Wagner, Volker Slowik, Gideon P.A.G. van Zijl, William P. Boshoff, Suvash C. Paul, Viktor Mechtcherine and Koichi Kobayashi

Abstract The durability of strain-hardening cement-based materials (SHCC) is strongly influenced by the transport of different substances through the material. Since numerous fine cracks are formed in SHCC, the relationship between the crack pattern and different transport properties has been the subject of experimental investigations, including tests of water and gas permeability, chloride ingress, and capillary absorption. It appears to be insufficient to consider only the average or maximum crack width when transport through SHCC is to be modelled. Therefore, the determination of certain crack pattern parameters has been proposed that take into account both crack widths and distances. These parameters may be linked to certain transport properties of the respective material.

Keywords Transport · Crack patterns · Permeability · Capillary absorption · Chloride ingress

2.1 Introduction

Cracks in concrete members are known to have a significant influence on the transport of fluids, gases and ions through cement-based materials and, thereby, on the structural durability of the respective members. For this reason, crack widths are

C. Wagner (✉) · V. Slowik
Leipzig University of Applied Sciences, Leipzig, Germany
e-mail: christian.wagner@htwk-leipzig.de

G.P.A.G. van Zijl · W.P. Boshoff · S.C. Paul
Stellenbosch University, Stellenbosch, South Africa

V. Mechtcherine
Technische Universität Dresden, Dresden, Germany

K. Kobayashi
Gifu University, Gifu, Japan

G.P.A.G. van Zijl and V. Slowik (eds.), *A Framework for Durability Design with Strain-Hardening Cement-Based Composites (SHCC)*, RILEM State-of-the-Art Reports 22,
DOI 10.1007/978-94-024-1013-6_2

normally restricted by design codes, as, for instance, to about 0.2 mm in ordinary steel-reinforced concrete members. Limits of such an order of magnitude are not acceptable for the fracture behaviour of strain-hardening cement-based materials (SHCC), which is characterised by the formation of a large number of individual cracks with a spacing of only a few millimetres. In order to ensure sufficient durability of such materials, the control and the characterisation of crack patterns appears to be an important goal of durability design.

Knowledge of the transport properties of uncracked SHCC may contribute to the adequate modelling of certain transport mechanisms, but it appears to be impossible to evaluate the durability of the material on the basis of these properties alone. This stems from the fact that transport through cracks tends, in general, to occur much faster than that which occurs through the uncracked material. Hence, the combined action of mechanical loading and exposure to media that might be transported through the cracked material requires consideration.

Most of the reported investigations into transport through cement-based materials have, to date, focused on water as the transport medium. The reason for this focus is that water may directly affect the material properties of the SHCC while also being the transport medium for ions which might be potentially damaging. Hence, the resistance of cracked SHCC to water transport should be maximised. This may be accomplished by controlling and restricting crack widths or by means of water-repellent treatment (see Wittmann et al. 2012).

Two physical processes are of importance for water transport through SHCC: permeability and capillary absorption. The effects of the cracks on these processes are different. Some typical experimental observations and proposed models for the two above-mentioned — and other — transport processes are explained further on in this chapter. The next section, however, is dedicated to the definition and determination of crack pattern parameters that will later be linked to the transport properties.

2.2 Crack Pattern Characterisation

The evaluation of test results that expectedly depend on crack patterns appears to be a challenging task. In repeated tests, it has proved to be impossible to reproduce identical crack patterns concerning crack density, crack width distribution, and crack spacing distribution. Furthermore, it has become apparent that neither the maximum crack width, nor the average crack width may be exclusively used for the evaluation of transport properties. The inability results from the non-proportional influence of the crack width and other crack pattern-related parameters on the transport properties. The strain level is also not directly connected to the transport properties, since the resulting crack pattern depends both on the material properties and on the particular loading conditions involved. Instead, it is proposed to base the estimation of the transport properties on the crack patterns actually observed at the surface of test specimens or structural members. The crack patterns should then be characterised by considering, in addition to the number of cracks, the crack width distribution and also, especially in the case of capillary absorption, the crack spacing distribution.

With respect to the durability of SHCC, the crack widths are of utmost importance. Their critical nature relies on them having a disproportionate influence on the transport rate for several transport mechanisms, as well as on them also affecting the self-healing potential of the cracks concerned. The number of cracks per unit length, which is a property that is also referred to as crack density, normally has a linear influence on the transport rate involved. In the case of capillary absorption, the individual crack distances also influence the transport rate, especially when the distances concerned are small. The uncracked material between neighbouring cracks absorbs water, but only until the space between them is fully saturated. Therefore, the crack spacing distribution has to be considered, in the case of capillary absorption, as being a third property of the crack pattern. The variation of the crack depths is not considered in the following discussion of crack pattern parameters, since the crack observation is currently almost exclusively undertaken at the surface of SHCC samples. For the durability of real structures, however, crack depths are highly significant.

Prior to the characterisation of crack patterns, they require to be documented, with the cracks involved requiring identification, and the measuring of their widths. This is usually accomplished by means of such non-contact observation methods as digital image processing (DIP), or digital image correlation (DIC). Adequate techniques are explained and compared by Van Zijl et al. (2016), who report on a comparative mechanical test series initiated by the RILEM (International Union of Laboratories and Experts in Construction Materials, Systems and Structures) Technical Committee 208-HFC, Subcommittee 2. As indicated in Chap. 1, five laboratories participated in the study concerned. The purpose of the tests conducted was to compare and discuss experimental observations in terms of crack widths and spacing, and to make recommendations for a suitable test set-up and procedure towards characterising crack patterns on SHCC surfaces.

The characterisation of crack patterns requires the derivation of certain quantitative measures, so as to allow for the making of comparisons and further evaluations. Boshoff and Adendorff (2010) proposed a crack pattern evaluation method with respect to the durability of SHCC, with the method in question considering both the crack width distribution and the number of cracks present. The method yields a single dimensionless numerical value of which the use was recommended in respect of the estimating of the damage potential of cracked SHCC. An advanced version of the method identified has been presented by Boshoff et al. (2016). It also yields single numerical values, characterising the ingress potential of different deleterious materials in cracked SHCC. Due to the high impact of the cracks on the transport mechanisms involved, the transport through the uncracked matrix has been neglected (Boshoff et al. 2016). First, the so-called crack ingress potential (*CIP*) is determined that defines the relative significance of the ingress into a single crack, due to a particular transport mechanism. The relationship between the *CIP* and the crack width has to be based on experimental investigations into the respective transport mechanism. Examples for the *CIP* are given by Boshoff et al. (2016) for water permeability, based on the experimental results of Wang et al. (1997), and for chloride diffusion, based on experimental results of Djerbi et al. (2008). The following equations apply.

For water flow:

$$CIP = 4.84 \cdot 10^{-17}\, w_c^{5.62} \tag{2.1}$$

For chloride diffusion:

$$CIP = \begin{cases} 0 & \text{for } w_c \leq 30\,\mu\text{m} \\ 4.115 \cdot 10^{-11} \cdot w_c^2 - 1.235 \cdot 10^{-15} \cdot w_c & \text{for } 30\,\mu\text{m} < w_c \leq 100\,\mu\text{m} \\ 2.881 \cdot 10^{-15} \cdot w_c & \text{for } w_c > 100\,\mu\text{m} \end{cases} \tag{2.2}$$

Expectedly, the *CIP* increases with the crack width w_c. The *CIP* is then used for weighting the crack width distribution. For the latter, a log-normal probability density function is adopted:

$$f(w_c) = \frac{1}{w_c\sqrt{2\pi}\sigma} e^{\frac{-(\ln(w_c)-\mu)^2}{2\sigma^2}} \tag{2.3}$$

with the parameters μ and σ.

The above yields the so-called *IPD*.

For water flow:

$$IPD = \frac{4.84 \cdot 10^{-17} \cdot w_c^{5.62}}{w_c\sqrt{2\pi}\sigma} e^{\frac{-(\ln(w_c)-\mu)^2}{2\sigma^2}} \tag{2.4}$$

For chloride diffusion:

$$IPD = \begin{cases} 0 & \text{for } w_c \leq 30\,\mu\text{m} \\ \frac{4.115 \cdot 10^{-11} \cdot w_c^2 - 1.235 \cdot 10^{-15} \cdot w_c}{w_c\sqrt{2\pi}\sigma} e^{\frac{-(\ln(w_c)-\mu)^2}{2\sigma^2}} & \text{for } 30\,\mu\text{m} < w_c \leq 100\,\mu\text{m} \\ \frac{2.881 \cdot 10^{-15} \cdot w_c}{w_c\sqrt{2\pi}\sigma} e^{\frac{-(\ln(w_c)-\mu)^2}{2\sigma^2}} & \text{for } w_c > 100\,\mu\text{m} \end{cases} \tag{2.5}$$

Finally, the *IPD* is integrated over the crack widths and multiplied by the so-called crack intensity C_i. This yields the Ingress Potential Index (*IPI*) as a single dimensionless numerical value:

$$IPI = C_i \int_0^{\infty} IPD\, dw_c \tag{2.6}$$

with $C_i = \frac{\text{number of cracks per metre}}{1000}$

Figure 2.1 shows, for a certain example, the unweighted and weighted probability density functions of the crack width at different strain levels, as well as the *IPI* for chloride diffusion and water flow (Boshoff et al. 2016). Due to the weighting by the *CIP* for water flow (see Fig. 2.1b), the larger cracks might be seen to have a more significant impact on the ingress potential than do the smaller ones. Figure 2.1c shows that the two transport mechanisms considered here exhibit different strain dependencies.

It has to be taken into account that the *IPI* values for different transport mechanisms cannot be directly compared because of the independent normalizations involved. Currently, the concept does not allow for the quantitative derivation of a flow rate, or of an absorbed water volume, for a certain crack pattern under a given exposure. In addition, the interaction between neighbouring cracks is not yet considered, with the crack spacing distribution not yet being an input parameter. In the case of water permeability, the above is not expected to cause significant error, but, in the case of capillary absorption, the individual crack distances are of importance (see Sect. 2.6).

Motivated by the work of Boshoff and Adendorff (2010), an alternative method of crack pattern evaluation was developed by Wagner (2016), Wagner and Slowik (2011), and Wagner et al. (2011, 2012). The main advantage of the newly

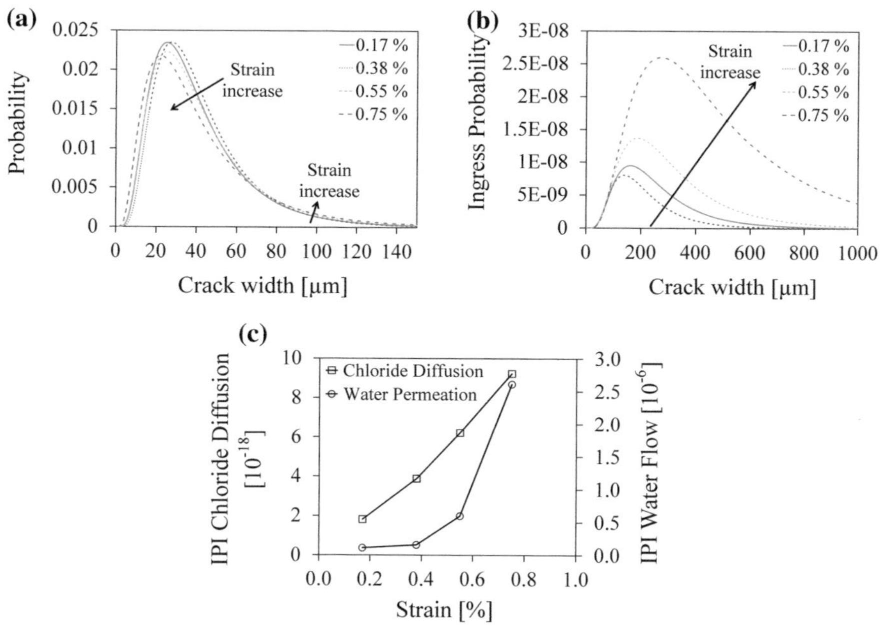

Fig. 2.1 **a** The unweighted and **b** weighted (*IPD* for water flow) probability density functions of the crack width at different strain levels, and **c** the Ingress Potential Index (*IPI*) for chloride diffusion and water flow (reproduced from Boshoff et al. 2016)

developed approach is the physically justified and quantitative linking of adequate crack pattern parameters to certain transport properties, for instance to the water flow in m^3/s, or to the capillary absorption in kg/m^2. Such linking enables a comparison to be drawn with allowable or acceptable values. Due to the different transport mechanisms, different crack pattern parameters are proposed for permeability and capillary absorption, respectively, although, in each case, they are expressed as a single numerical value.

As a prerequisite, it is necessary to present the measured crack width distributions as so-called crack width polygons (*CWP*) (see Fig. 2.2). The format of the curves resembles that of grain-size distribution curves. For the respective crack pattern, the percentage of cracks with certain widths is cumulated. This method of presentation allows for the quick determination of the percentage of cracks, the width of which does not exceed a certain limit. The residual strain, as indicated in Fig. 2.2, is the sum of the crack widths in the unloaded specimen related to the evaluation length. Deformations of the material between the cracks are not considered. Note that the *CWP* is completely independent of the crack density, i.e., of the number of cracks.

On the basis of the *CWP*, the so-called crack width value (*CWV*) may be determined by adding the complements of the frequency values for all crack width classes i:

$$CWV = \frac{\sum_{i=0}^{n} (100 - CWP(w_i))}{100} \tag{2.7}$$

It has to be taken into account that the *CWV* depends on the class width Δw of the crack width distribution. For a constant Δw, the product $(CWV \cdot \Delta w)$ yields the average crack width. The major advantage of assigning the individual cracks to certain classes and of deriving the *CWV* is the opportunity to weight each class according to certain transport mechanisms (see Sect. 2.3). In previous publications by Wagner and Slowik (2011) and Wagner et al. (2011, 2012), the symbol *CCV* (for cumulated crack width value) has been used, instead of *CWV*. The reason for

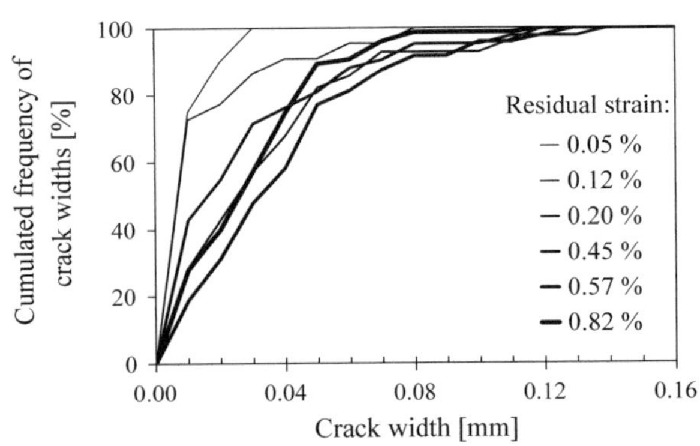

Fig. 2.2 Crack width polygons (*CWP*) for six damage levels, as expressed by the residual strain (Wagner and Slowik 2011)

the name change was the intended consistency of the notation for different crack pattern parameters.

In an analogy in reference to the crack width value (*CWV*), a crack spacing value (*CSV*) and a crack length value (*CLV*) may be determined.

$$CSV = \frac{\sum_{i=0}^{n}(100 - CSP(w_i))}{100} \quad (2.8)$$

$$CLV = \frac{\sum_{i=0}^{n}(100 - CLP(w_i))}{100} \quad (2.9)$$

with: *CSP* the crack space polygon
CLP the crack length polygon

The crack density (*CD*), meaning the number of cracks per unit length, also requires consideration when evaluating crack patterns with respect to the transport properties involved. Table 2.1 shows examples of the crack pattern characterisation by the proposed method. Assuming that the observed cracks completely split the specimens concerned, the *CLV* is constant for each series and, when multiplied by the class width Δl, it yields the specimen width. The values in the three lines at the bottom of Table 2.1, which are required for modelling water permeability, will be explained in Sect. 2.3.

2.3 Water Permeability

The pressure gradient-induced water transport in SHCC is dominated by the flow through the cracks, with the contribution of the material between the cracks being of minor importance. Such dominance stems from the comparatively high flow rate through the cracks, which can be described in terms of the Hagen–Poiseuille equation:

$$Q_t = \frac{\Delta p \cdot w^3 \cdot l}{d \cdot 12 \cdot \eta} \quad (2.10)$$

with:

- Q_t theoretical flow rate (m^3/s)
- Δp pressure difference (Pa)
- w crack width (m)
- l crack length (m)
- d specimen thickness (m)
- η dynamic viscosity (Pa·s)

The theoretical flow rate Q_t is a maximum value that is based on the assumption of laminar flow between smooth parallel-sided and continuous planes. The value

Table 2.1 Crack pattern parameters of selected samples and the corresponding normalised theoretical water flow rates

					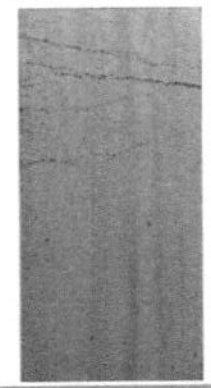	
Series	1	4	1	4	1	4
Residual strain (%)	0.12	0.10	0.48	0.45	0.65	0.60
CD (cracks/m)	35	55	111	83	99	189
CWV (−) $\Delta w = 0.01$ mm	2.0	2.3	4.8	6.0	7.0	3.8
CSV (−) $\Delta s = 1.0$ mm	9.1	15.2	8.5	10.9	9.4	5.4
CLV (−) $\Delta l = 10$ mm	4.0	6.0	4.0	6.0	4.0	6.0
CWV_{WP} (−)	264	61	292	639	2077	123
R_{WP} (mm^3•cracks/m)	0.0091	0.0034	0.0326	0.0533	0.2056	0.0233
$Q_{t,normalised}$ (−)	1.00	0.36	3.56	5.71	22.48	2.50

concerned is unreachable in real materials. The influence of all flow-reducing effects is considered in terms of a flow coefficient ξ_{WP} that relates the measured flow rate Q_m to the theoretical flow rate Q_t.

$$\xi_{WP} = \frac{Q_m}{Q_t} \tag{2.11}$$

Note that the proportionality of the theoretical flow rate to the third power of the crack width points to the importance of limiting crack widths in the durability design of SHCC members. When, for example, a crack of width 0.2 mm is replaced by four fine cracks of width 0.05 mm, the theoretical flow rate is reduced by 93.75%.

Experimental investigations into the water permeability of SHCC were reported by Lepech and Li (2009). The measured flow rates were related to the maximum crack width, and to the tensile strain. The average crack width did not exceed 70 µm, up to a tensile strain of 3%. Reference permeability tests with steel-reinforced mortar revealed that the SHCC had a six orders of magnitude smaller water permeability.

Mechtcherine and Lieboldt (2011) conducted permeability tests of cracked textile-reinforced concrete. Concerning the crack pattern created under tension, similarities exist between such a type of material and SHCC. It could be shown that the transport rate overproportionally increases along with the tensile strain. This observation may be attributed to the increasing crack widths involved. The measured flow rate was compared to a calculated one based on the Hagen–Poiseuille equation, whereby the flow coefficient ξ varied between $0.28 \cdot 10^{-3}$ and $0.75 \cdot 10^{-3}$. Furthermore, it was observed that the self-healing of fine cracks significantly reduced the water permeability concerned over time.

Wagner (2016) conducted water permeability tests of pre-cracked SHCC specimens. The specimens were cut out of preloaded dumbbell-shaped specimens of a length of 310 mm, with the inner part having a length of 100 mm, and a constant cross-section (see Fig. 2.3). The specimens were subjected to uniaxial tension up to certain strain levels (0.25, 0.50, 1.00, 1.50%). Thereafter, the specimens were unloaded, with the inner part then being used for permeability and capillary absorption tests.

Prior to the permeability tests, the crack patterns on the front and back of the unloaded specimens were photographically documented and evaluated. Thereafter, the side faces of the specimens were sealed, so as to allow for one-dimensional transportation during the permeability tests. When 28 days old, the unloaded specimens were, first, water stored and then, on one side, subjected to water pressure. In previous permeability tests of cracked steel-reinforced concrete, such as those published by Edvardsen (1996), pressure gradients of up to 0.5 MPa/m were applied. Such a value is realistic for the usual concrete members with thicknesses of at least 25 cm. In thin repair layers, however, significantly higher pressure gradients of more than 5 MPa/m can occur. Mechtcherine and Lieboldt (2011) performed permeability tests with pressure gradients of the same order of magnitude (i.e.

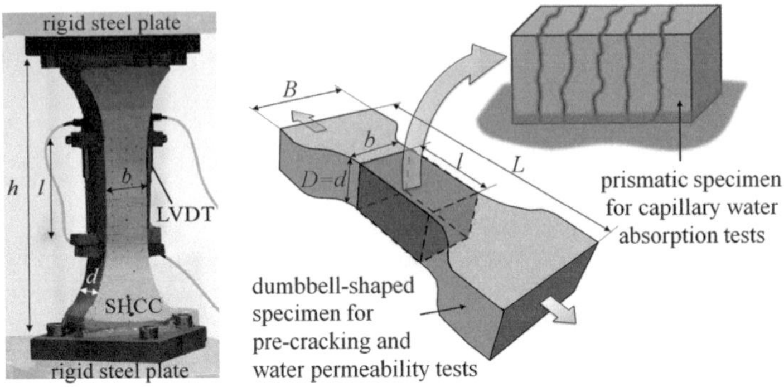

Fig. 2.3 Pre-cracking of specimens for permeability and capillary absorption tests (Wagner 2016)

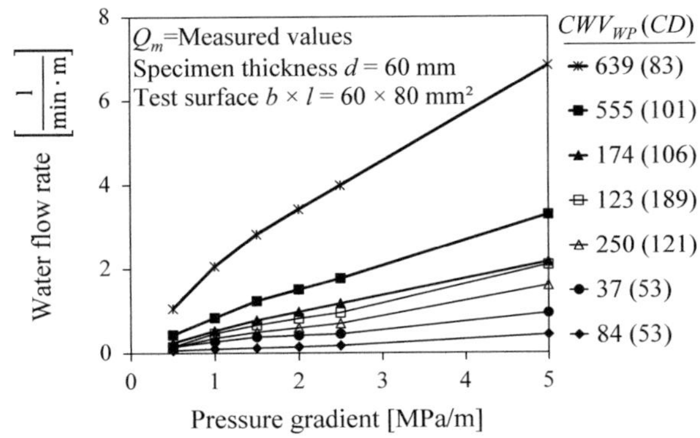

Fig. 2.4 Water flow rate, as dependent on the pressure gradient (Wagner 2016)

12.5 MPa/m) in cracked textile-reinforced concrete. In the experiments described in the current chapter, the applied pressure gradients ranged from 0.5 to 5 MPa/m. In addition to the pressure gradient and to the damage level expressed by the residual strain after pre-cracking, the specimen thickness was also varied. Figure 2.4 presents the results of one particular test series (Wagner 2016). The figure shows the water flow rate versus the pressure gradient for different damage levels, i.e., for different residual strains after pre-cracking and unloading. Each curve represents the behaviour of one particular specimen, with its unique crack pattern. The water flow rates in l/(min·m) presented in the diagrams in the present section are given per unit width, with them having been normalised to the same specimen thickness.

The evaluation of the results presented in Fig. 2.4 revealed a problem associated with the strain values used for characterising the damage level. As no clear dependency between flow rate and residual tensile strain could be identified, it appeared to be necessary to relate the water flow rate not to a strain value, but to certain crack pattern parameters. Such an approach will be followed later on in the present section.

In order to provide test conditions that realistically resembled real SHCC applications, Wagner (2016) conducted additional tests at SHCC overlays on concrete substrate with artificial crack (see Fig. 2.5). For creating cracks in the SHCC overlays, so-called zero-span elongation tests were conducted (Fig. 2.5a). The respective specimens consisted, in each case, of two concrete cuboids ($70 \times 70 \times 75$ mm^3) that were placed on top of each other, with the joint between the cuboids serving as an artificial crack. By means of displacing the cuboids, the artificial crack could be opened, or closed. The SHCC layers, which had been applied to two opposite side faces of the cuboids in question, were locally stressed when the artificial crack was opened. In Fig. 2.5a it can be seen that the wide-opened artificial crack between the cuboids did not 'break through' the SHCC overlay. The latter is also cracked, but with numerous fine cracks, rather than with a single crack.

The principle of the permeability tests at the pre-cracked SHCC overlays is shown schematically in Fig. 2.5b. In the loaded state, spacers had been placed between the two concrete cuboids so as to keep the artificial crack open during the subsequent permeability test. Despite the intended symmetrical preloading, different crack patterns were formed in the overlays on the two opposite sides of the concrete cuboids serving as substrate. Therefore, the crack patterns were documented and characterised on both surfaces. The parallel side faces of the samples were sealed so as to force the flow through one of the SHCC overlays and part of the artificial crack. Figure 2.6 shows the measured water flow rate versus the pressure gradient for an SHCC layer with a thickness of 20 mm. The individual curves, corresponding, in terms of their crack patterns, to certain specimens, exhibit a deviation from proportionality that is contradictory to the prediction of the Hagen–Poiseuille equation. This may be attributed to a deviation from laminar flow conditions, as

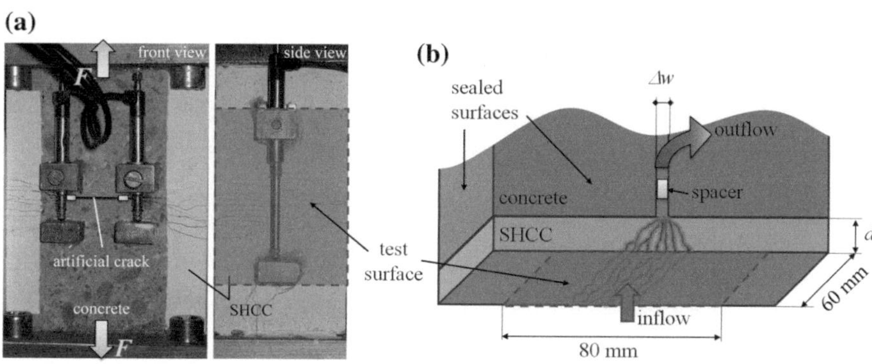

Fig. 2.5 **a** Pre-cracking of the SHCC overlay, and **b** the permeability test (Wagner 2016)

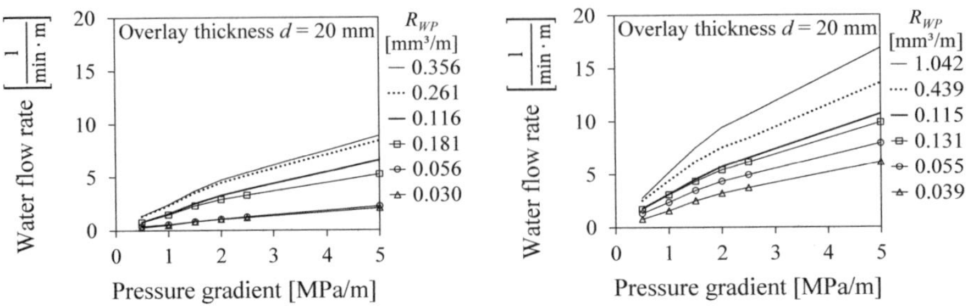

Fig. 2.6 Water flow rate, as dependent on the pressure gradient for an SHCC layer thickness of 20 mm (Wagner 2016)

well as to other flow-reducing effects, as, for instance, the particle plugging of cracks. It was also found that the flow rate appears not to be directly related to an overall strain value, which, in the present case, would be dependent on the artificial crack opening. Again, this observation points to the necessity to relate the permeability to the actual crack pattern parameters involved.

In the following, a method is proposed that links the water permeability to the crack pattern (Wagner 2016; Wagner and Slowik 2011). The pressure gradient-induced water flow rate depends on the crack density (*CD*), specifically on the number of cracks per unit length, and on the crack width distribution. The latter is highly relevant, due to the above-mentioned predominant effect of the crack width on the water flow rate. It can be expressed by means of the crack width value (*CWV*) calculated on the basis of the crack width polygon (*CWP*). However, it is necessary to weight each class of crack widths with respect to its influence on the water permeability. An adequate weighting function for this particular transport mechanism $W_{WP}(i)$ has been derived on the basis of the Hagen–Poiseuille equation (Wagner 2016):

$$W_{WP}(i) = 3i^2 + 3i + 1 \tag{2.12}$$

Using the function stated, the crack width value CWV_{WP} weighted for water permeability is obtained, as follows:

$$CWV_{WP} = \frac{\sum_{i=0}^{n} W_{WP}(i) \cdot (100 - CWP(w_i))}{100} \tag{2.13}$$

The *CWV* has to be considered as generally depending on the chosen class width Δw.

The theoretical water flow rate Q_t may now be calculated as follows:

$$Q_t = \frac{\Delta p \cdot b \cdot l \cdot \Delta w^3}{d \cdot 12 \cdot \eta} \cdot CWV_{WP} \cdot CD \tag{2.14}$$

with:

b	evaluation length
Δw	chosen class width for the *CWP*
CWV_{WP}	weighted crack width value
CD	crack density

All other symbols were introduced with the provision of the Hagen–Poiseuille equation (see above). The product $b \cdot CD$ yields the total number of cracks. All crack pattern dependent influences on the theoretical flow rate Q_t may be combined and expressed by means of a so-called hydraulic crack pattern parameter R_{WP}. Doing so facilitates the evaluation of observed crack patterns, as well as their comparison with respect to structural durability.

$$R_{WP} = \Delta w^3 \cdot CD \cdot CWV_{WP} \tag{2.15}$$

On the basis of the hydraulic crack pattern parameter R_{WP}, the flow rate may be predicted more reliably than on the basis of the irreversible strain. This is demonstrated in Fig. 2.7, in which the flow rate that was measured by Wagner (2016) in uniaxially pre-cracked specimens, is plotted versus the irreversible strain (Fig. 2.7a), and versus the hydraulic crack pattern parameter R_{WP} (Fig. 2.7b). In the last-mentioned case, a better correlation is obtained than in the previous one.

Corresponding results for the SHCC overlays on concrete substrate with artificial crack are presented in Fig. 2.8. As in the previous case, a better correlation is obtained if the flow rate is plotted versus the hydraulic crack pattern parameter R_{WP}.

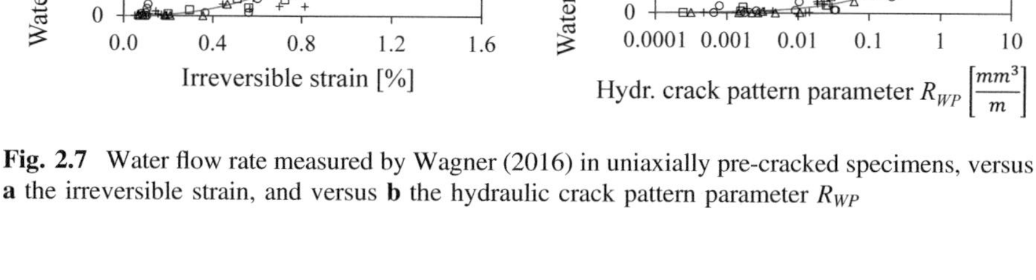

Fig. 2.7 Water flow rate measured by Wagner (2016) in uniaxially pre-cracked specimens, versus **a** the irreversible strain, and versus **b** the hydraulic crack pattern parameter R_{WP}

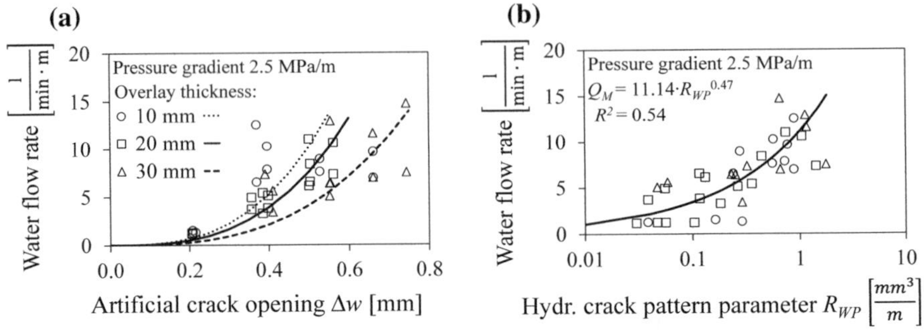

Fig. 2.8 Water flow rate measured by Wagner (2016) at the SHCC overlays, versus **a** the artificial crack opening, and versus **b** the hydraulic crack pattern parameter R_{WP}

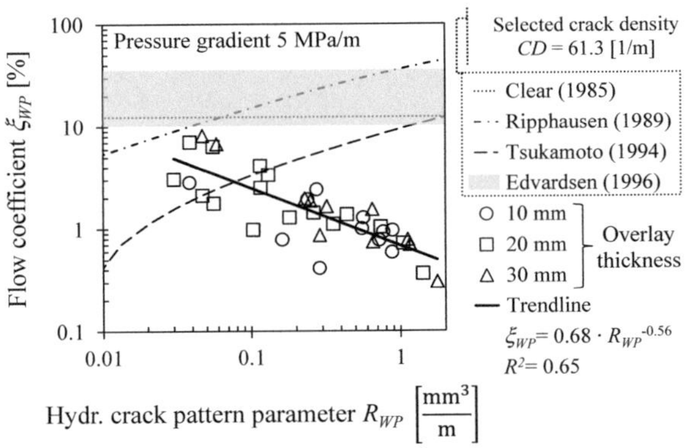

Fig. 2.9 Flow coefficient ξ_{WP} versus the hydraulic crack pattern parameter R_{WP} with Wagner's (2016) experimental results being compared to those of other authors

In Fig. 2.8a, the influence of the overlay thickness can be seen. Functions $Q = A \cdot w^3$ were fitted to the measured data. Expectedly, a clear trend of reduced permeability for thicker overlays exists, despite the comparatively large scatter when the artificial crack opening serves as abscissa.

Figure 2.9 shows the flow coefficient ξ_{WP} as being dependent on the hydraulic crack pattern parameter R_{WP}. As explained before, this coefficient is the ratio of measured and theoretical flow rate. Wagner's (2016) experimental results are compared to the results that have been reported by other authors for different cement-based materials. The opposite trend found by Wagner (2016) represents a material-specific behaviour of SHCC that is mainly attributed to the characteristic crack width distribution, which is expressed in the given instance by means of the weighted crack width value CWV_{WP}.

2.4 Gas Permeability

Very little information is available on the air permeability of cracked SHCC. Mechtcherine et al. (2007) reported on permeation tests performed on uncracked and cracked SHCC specimens (Ø 100 mm, height of 60 mm) cored from dumbbell-shaped prisms. Figure 2.10 shows the cracked cored specimens and a schematic view of the test set-up used for the air permeability measurements.

The obtained results indicate that the air permeability of uncracked SHCC is within the range of typical values for ordinary concrete. When the material is loaded up to a tensile strain of 0.5%, an increase of almost two orders of magnitude has been observed (Mechtcherine et al. 2007) (see Fig. 2.11). Whereas such values are

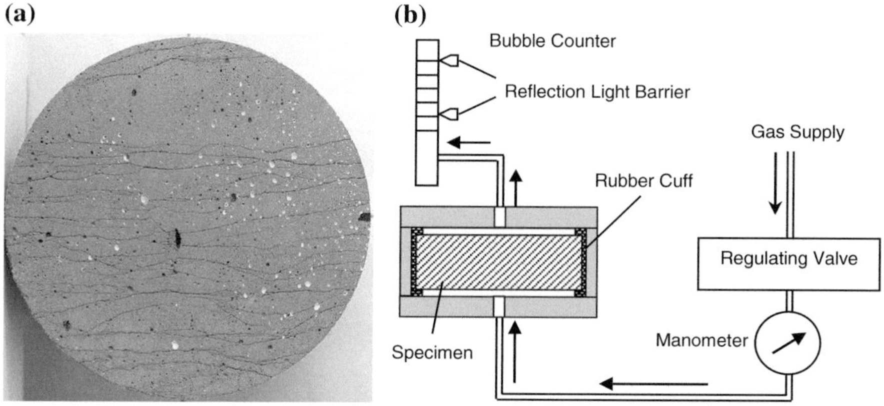

Fig. 2.10 **a** Cylindrical test specimen (cracked) cored from prism tested to failure under monotonic loading (ε = 6.5%), **b** schematic view of the test set-up used for air permeability measurements

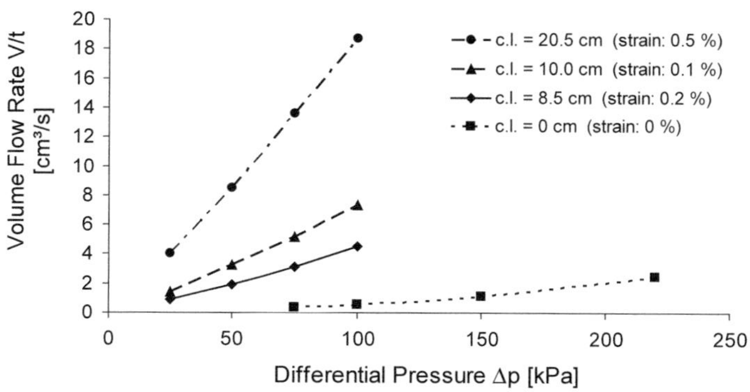

Fig. 2.11 Air volume flow rate as a function of differential pressure for selected specimens with different residual strain levels, with c.l. being the cumulative crack length (Mechtcherine et al. 2007)

no longer comparable to those of sound concrete, it should be noted that the specimens were tested at an age of 45–50 days, and that the mix had a high content of fly ash. Thus, it can be expected that, in field applications with similar mixes, ongoing hydration and self-healing should have a mitigating influence over time.

Whereas all crack widths present were found to be smaller than 0.1 mm, their number varied significantly and generally increased with increasing strain level. To quantify the pre-damage, the cumulative crack length was measured for each specimen. Only pass-through cracks were considered, meaning cracks going completely through the specimen. Figure 2.12 shows all results of the air volume flow measurements at a pressure difference of 50 kPa. A pronounced linear correlation between volume flow rate and cumulative crack length was found for the considered strain range from 0 to 1%.

Mechtcherine and Lieboldt (2011) reported on gas permeability tests of cracked textile-reinforced concrete. As previously stated, the crack patterns created in the

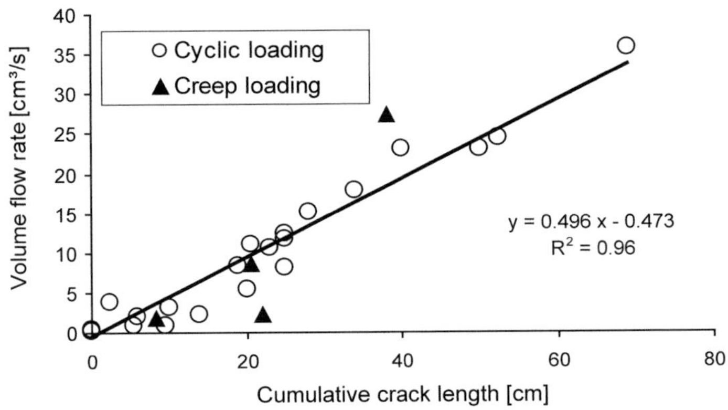

Fig. 2.12 Effect of the cumulative crack length on the air volume flow rate, cylinders with Ø 100 mm and height of 60 mm, differential pressure 50 kPa, room temperature (Mechtcherine et al. 2007)

Fig. 2.13 Oxygen flow rate in cracked textile-reinforced concrete under increasing tensile strain, at a pressure difference of 3 kPa (Mechtcherine and Lieboldt 2011)

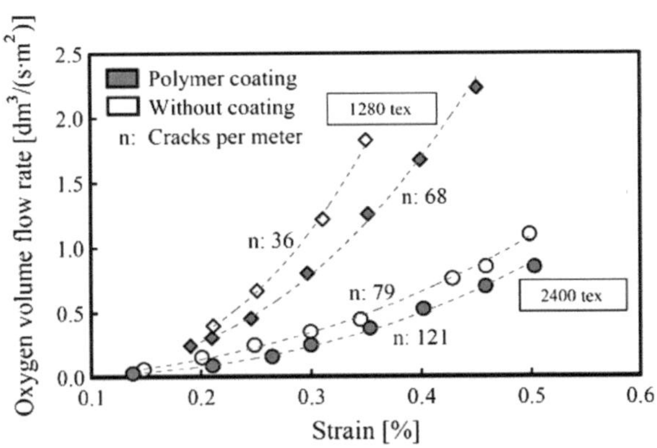

type of material mentioned are similar to those found in SHCC. Figure 2.13 shows the oxygen flow rate measured in a permeability cell under uniaxial tension at a pressure difference of 3 kPa. Under increasing strain, the crack widths can be seen to be growing larger, resulting in the hyperproportional increase of the flow rate. For the uncoated yarn of a fineness of 1280 tex (see the upper curve in Fig. 2.13), the average crack width in the unloaded state amounted to 65 μm, which is in the order of magnitude of what is expected for SHCC. The other curves correspond to smaller average crack widths.

2.5 Chloride Ingress

Steel corrosion in reinforced concrete (RC) is known to be caused by carbonation or chloride, with models having been proposed for both deterioration mechanisms (DuraCrete 2000). Chloride-induced corrosion in steel-reinforced SHCC (R/SHCC) is described in more detail in Chap. 9. In the current section, chloride ingress is reported for both reinforced and unreinforced SHCC.

2.5.1 Chloride Profile in Cracked and Uncracked SHCC

Kojima et al. (2014) immersed an entire SHCC test specimen, which was pre-cracked, in a 3% NaCl solution for a period of one month, and then, after breaking the test specimen apart, analysed the chloride content in the fracture surface. The results showed that 4 kg/m^3 or more of chloride per unit volume of mortar (i.e. approximately 0.6% or more of the cement mass) had been accumulated in the fracture surface. Such quantity may have caused embedded steel to corrode, if a sufficient amount of oxygen was present. In the case of the penetration of chloride aqueous solution to the interior of cracks, however, the results may differ, depending on the conditions under which the liquid was provided: by spraying; by alternate wetting and drying; or by immersion.

Chloride profiles determined by chemical titration are shown in Fig. 2.14 (Wittmann et al. 2011). Dumbbell specimens were first loaded in a stiff steel test rig. Then, chloride penetration into the specimens under imposed strain was determined by means of a modified Karsten tube. In this case, a surface of 40 × 100 mm^2 was put in contact with a box containing 3.1% NaCl solution. After the contact between the SHCC and the salt solution had been maintained for three hours, thin layers were removed successively from the surface, which had been in contact with the salt solution, by means of milling. The free chloride content of the powder obtained was then determined by titration, according to RILEM TC 178-TMC (2002b). The dumbbell specimens were loaded to predetermined average tensile strains ranging from 0% (uncracked) to 2%. Higher free chloride content was found in the specimens that were loaded to higher average tensile strains. In the cracked specimens,

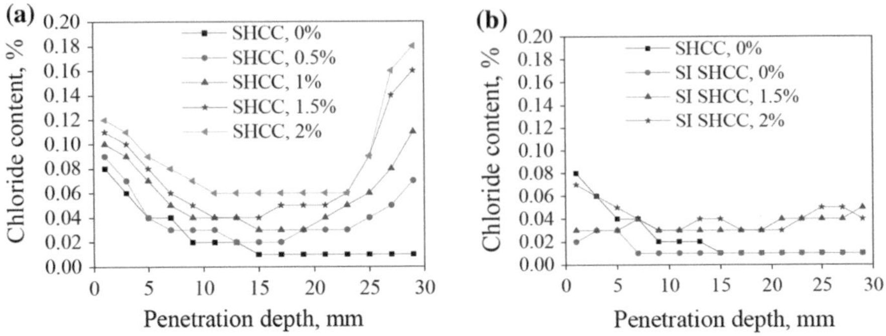

Fig. 2.14 Influence of imposed strain on free chloride profiles in unreinforced, uncracked and cracked SHCC, after contact of one surface with a 3.1% NaCl solution for three hours, showing specimens that were **a** untreated and **b** surface-impregnated with silane emulsion for the purpose of water repellence (Wittmann et al. 2011)

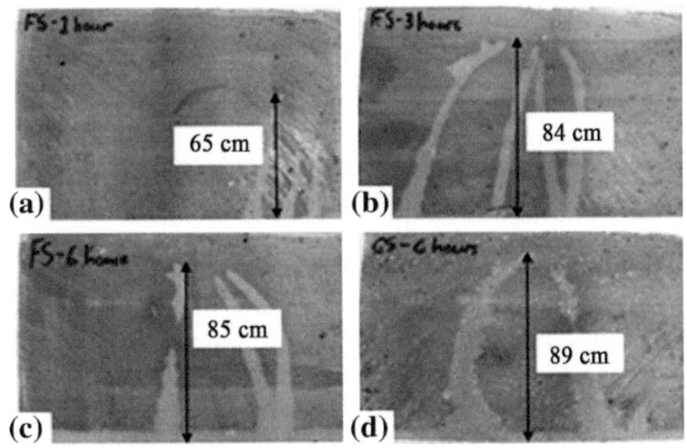

Fig. 2.15 Chloride penetration in 100 mm high, pre-cracked specimens after chloride exposure for **a** 1 h, **b** 3 h, **c** 6 h, and **d** a pre-cracked specimen containing course sand, after chloride exposure for 6 h (Paul and Van Zijl 2014)

significantly higher chloride content was seen at the opposite face of the immersed surface of the specimens, which was ascribed to evaporation at the opposite face causing significant deposits of chloride close to the face concerned (Wittmann et al. 2011). Figure 2.15b shows the free chloride profiles for similar SHCC specimens that were surface-treated with silane emulsion for water repellence.

Typically, in cracked SHCC specimens, the chloride penetration is rapid. In similar capillary absorption tests performed by Paul and Van Zijl (2014) on SHCC specimens that were pre-cracked in bending, chloride reached the full crack depth within one to three hours and, subsequently, the chloride penetrated from the crack faces into the matrix, as is shown in Fig. 2.15. The grey-white coloured areas show the areas to which chloride had penetrated, as prepared by silver nitrate spraying of the inner surface, which was exposed by cutting the specimen after the chloride absorption had taken place. The width of the cracks in the unreinforced SHCC was in the range of 50–100 μm. As the figure shows, in one to 6 h after the bottom of the test specimen had been immersed in a 3.5% NaCl solution, the chloride had penetrated to a depth of several centimetres along the cracks.

The total chloride content (% by weight of binder) profile was also determined for cracked steel-reinforced SHCC specimens subjected to cyclic wetting (three days) and drying (four days) exposure to 3.5% NaCl solution by means of capillary absorption and ponding, respectively (Paul 2015). Cyclic wetting and drying exposure to NaCl solution allows for the relatively deep penetration of chloride ions (Moukwa 1990), and it can lead to corrosion rates in RC that are 20 times higher than are those resulting from exposure to a continuous salt fog (Yeomans 2004). The material composition for the SHCC specimens tested by Paul (2015) is shown in Table 2.2. FS2, CS2, and FS31 specimens (nominally $100 \times 100 \times 500$ mm^3) were pre-cracked in three-point bending. In Series 1 (FS2, CS2), the specimens were unloaded after flexural cracking and subsequently exposed to cyclic wetting by means of capillary absorption and drying, meaning in the unloaded condition. In Series 2 (FS31), the specimens were also cracked in three-point bending, but then retained in special steel frames and subjected to cyclic ponding of the salt solution, meaning in the loaded condition. It should be noted that the specimens were loaded either to, or close to, their maximum capacity, so as to maximise crack widths for the current study. The full details of the crack widths that were recorded on the farthest tensile face are given in Paul (2015). The surface crack widths ranged between 0.02 and 0.25 mm, but with average widths in the range of 0.02–0.06 mm for the unloaded specimens (Series 1, FS2 and CS2), and between 0.02 and 0.25 mm applying in the case of the loaded specimens (Series 2, FS31), with their average crack widths being in the range of 0.03–0.17 mm. All the cracks formed in a central 200 mm long zone of the 500 mm long specimens. Series 1 was subjected to 88 weeks of cyclic absorption, with only the 200 mm central, cracked face being exposed to salt solution. In contrast, Series 2 was subjected to 37 weeks (with some up to 57 weeks) of cyclic ponding on the cracked central 200 mm length.

The chloride profiles shown in Fig. 2.16 for the cracked steel-reinforced SHCC specimens were determined by means of the X-ray fluorescence (XRF) method. In the present case, the powder samples were collected by means of drilling from the exposed surface in layers of 3 mm each, down to a depth of 45 mm in the specimens. A drill with diameter of 16 mm was used at four to six cracked positions to ensure that there was sufficient powder per layer. So, the chloride content obtained in the instance mentioned is the average value of different cracked positions within

Table 2.2 Amount of materials (kg/m^3) used in SHCC mixes (Paul 2015)

Type	Id	Cement	Fly ash	Sand	Water	PVA fibre
SHCC	Series 1: FS2, CS2[a]	392	674	553	392	2%
	Series 2: FS31					
	Series 3: FS32[b]					

[a]Note that the maximum aggregate size in the present case was 1.7 mm, whereas in others it was 0.25 mm

[b]Series 3 is discussed in Sect. 9.3

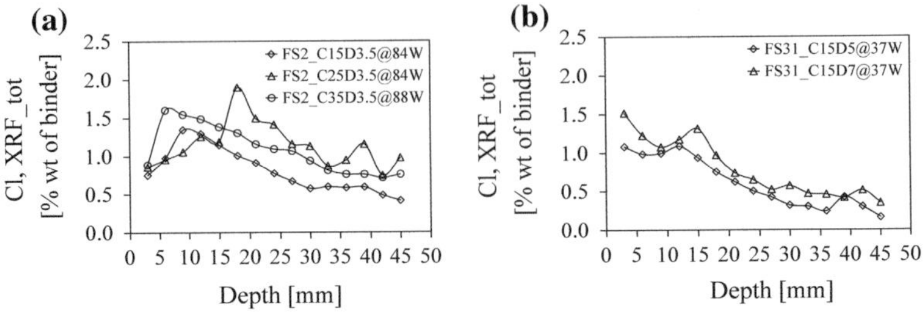

Fig. 2.16 Total chloride content determined by XRF after different periods of **a** capillary absorption and **b** ponding exposures in cracked steel-reinforced SHCC specimens (reproduced from Paul et al. 2016)

a single layer. Note further that different cover depths of the steel (15, 25 and 35 mm, as denoted by C15, C25, C35 in Fig. 2.16) were used, and that the bending tests were performed up to different deflection levels (3.5, 5 and 7 mm, as denoted by D3.5, D5, D7 in Fig. 2.16). The chloride profiling was done after a varying number of weeks, ranging from 37 weeks up to 88 weeks, as seen in the legend of Fig. 2.16. From Fig. 2.16, it is clear that significantly (an order of magnitude) higher chloride content was found than in cracked specimens of Fig. 2.14 that were subjected to absorption for only three hours.

Kobayashi et al. (2010) reported chloride penetration, as can be seen in Fig. 2.17, in cracked RC beams (of size 40 × 100 × 1800 mm^3) in which cracking

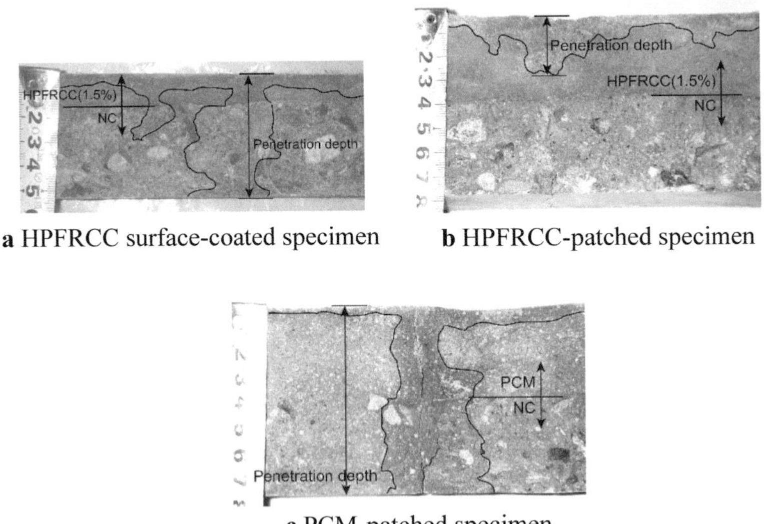

Fig. 2.17 Examples of chloride penetration profiles in HPFRCC specimens (Kobayashi et al. 2010)

was introduced by uniaxial tensile loading, up to a crack width of about 0.40 mm. In the present case, two steel reinforcing bars were used in the beams, with a cover depth of 25 mm, and with the 150 mm long specimens being cut out from the cracked part of the 1800 mm long beams. Before forming the cracks, the surface of the RC specimens was treated in three different ways: (a) a 10 mm layer of high-performance fibre-reinforced cement composite (HPFRCC) above the steel bars was applied to act as a surface coating to the specimen; (b) a 35 mm layer of HPFRCC was used to cover the steel bars of the patched specimen; and (c) a 35 mm layer of polymer cement mortar (PCM) was overlaid on the RC surface covering the steel bars of the patched specimen. The HPFRCC was reinforced with high-strength polyethylene (PE) fibres. The volumetric fibre content ranged from 0.75 to 1.5%, with only results for 1.5% being presented in Fig. 2.17. For chloride penetration, a 3% sodium chloride aqueous solution was sprayed for a period of 60 days. The spraying was conducted for a period of 5 min every 6 h. The coloured areas indicate the areas where chloride ions penetrated, as visualised by silver nitrate spraying. A comparison of Figs. 2.15 and 2.17 clearly shows that (i) the chloride penetration depth in SHCC at locations other than cracked locations was small; (ii) in test specimens (a) and (c), as shown in Fig. 2.17a, c, chloride penetrated to the interior through the cracks; and (iii) in the concrete, the chloride penetration depth from the cracks in the normal direction to the cracks is relatively large. SHCC, used as patch repair (Fig. 2.17b), showed excellent reinforcement corrosion-resistant performance in the tests undertaken in this regard, as will be discussed in Chap. 9.

For allowing strain-hardening behaviour based on the properties of PE fibres, the mortar matrix that was used in the study by Kobayashi et al. (2010) had a high strength, with a low water–cement ratio (w/c = 0.3). In contrast, the fibres used in the study by Paul (2015) were made of polyvinyl alcohol (PVA), and the w/c ratio was higher than the ratio mentioned above. In order to achieve pull-out of the fibres concerned under tensile stress, although the mortar matrix had a large powder content, a large quantity of fly ash was used in the mix, resulting in a w/c of 1.0 and a w/b of 0.37, if all the fly ash involved were considered as binder.

2.5.2 Total and Free Chloride Content in SHCC

The total and free chloride content in SHCC was determined by Paul (2015) by means of chemical titration, according to RILEM TC 178-TMC (2002a, b) recommendations. It is argued that not all the chloride in concrete contributes to the corrosion process. Some of the chloride binds with the paste present. The unbound, or free, chloride in concrete is believed to participate in the corrosion process. However, some of the bound chloride may dissolve over time and also participate in the corrosion process. This complex process involves many factors, such as the temperature, the relative humidity, the moisture content, the amount of binder content, and the microstructure of the concrete. For the particular mix design of

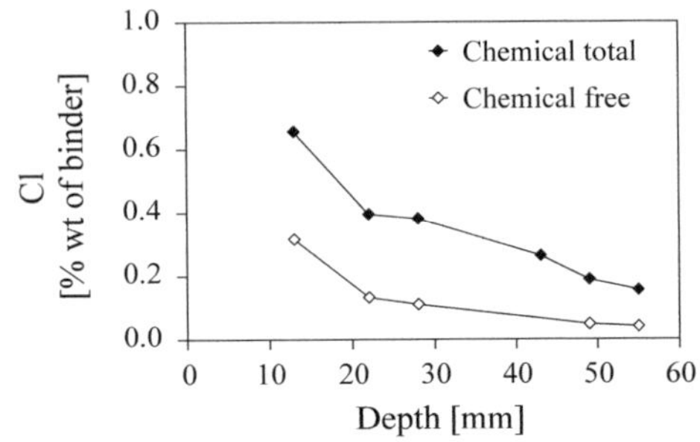

Fig. 2.18 Total and free chloride content in a precracked specimen of Series 3 (Table 2.2), after seven days of continuous ponding (reproduced from Paul et al. 2016)

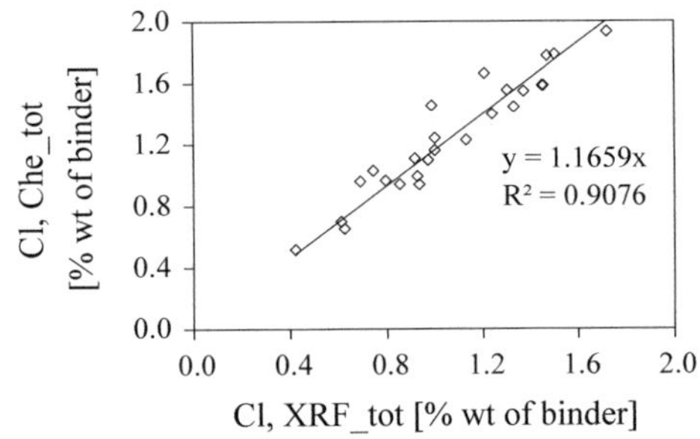

Fig. 2.19 Relationship between the XRF and the chemical analysis of total chloride content (reproduced from Paul 2015)

SHCC that is shown in Table 2.2, the free and total chloride contents from the chemical analysis are shown in Fig. 2.18. The ratio of free to total chloride content was found to be smaller at greater depths in the cracked reinforced SHCC.

Figure 2.19 shows the correlation between the total amount of chloride that was obtained by means of chemical analysis and that which was determined by XRF analysis for the same samples, as taken from the chloride-exposed reinforced SHCC specimens (Paul 2015). The strong, direct relationship indicates that XRF total chloride analysis may be an alternative to the chemical analysis concerned. The chemical analysis is significantly more time-consuming than is XRF.

2.5.3 Influence of Crack Patterns on Chloride Content in SHCC

The influence of the average surface crack width (ACW) on chloride penetration in reinforced SHCC was observed by Paul (2015), as is shown in Fig. 2.20. No correlation is seen to exist between crack width and chloride content in Fig. 2.20a. In Fig. 2.20b, it appears that the total amount of chloride increases as the ACW increases from 0.03 to 0.16 mm. Note, however, that although the ACWs are small in Fig. 2.20a, the specimens involved were exposed for a significantly longer period of time (for 108 weeks in total) than were those in Fig. 2.20b (for 57 weeks),

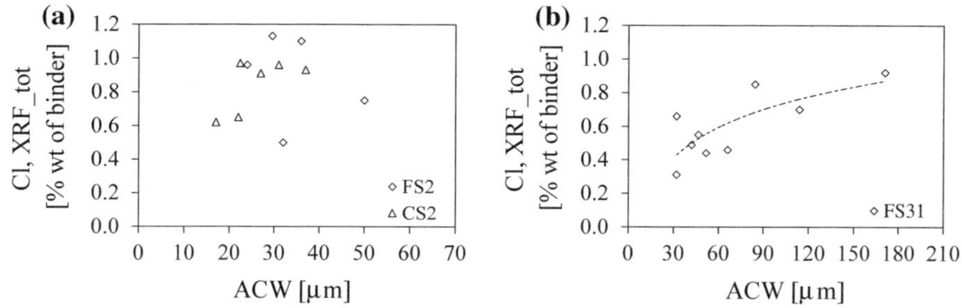

Fig. 2.20 XRF total chloride (Cl) and average crack widths (ACW) relationship in reinforced SHCC specimens, after **a** 108 weeks (Series 1), and after **b** 57 weeks (Series 2) of chloride exposure (reproduced from Paul 2015)

allowing the migration of chloride not only into the cracks, but also, subsequently, into the uncracked matrix between the cracks. Chloride from both the cracked and the uncracked parts were collected by means of drilling with a large (16 mm) diameter drill for the sampling and chloride profiling that was performed by Paul (2015).

The chloride diffusion Ingress Potential Index (IPI_{Cl}) that was derived from crack width distributions and chloride diffusion in cracks, as elaborated on in Sect. 2.2, as well as the unweighted *CWV*, which was also described in Sect. 2.2, were determined from the crack pattern data reported by Paul (2015) for the specimens of Series 1 and 2. Note that the chloride profiles were determined for twelve specimens in total, three specimens of Series 1 (FS2) and nine specimens of Series 2 (FS31), of which six after 37 weeks of cyclic chloride ponding and three after 57 weeks. The results obtained are presented in Fig. 2.21. From the graphs shown, a reasonable correlation was found with the total chloride content at the relatively low duration of exposure (37 weeks) for both of the indices involved (IPI_{Cl} and *CWV*). For longer exposure periods than the above, chloride has been found to continue to accumulate in the specimens. Such holds true even for the specimens of Series 1 that were exposed in the unloaded state, and which, thus, had relatively fine crack widths, as expressed by their lower IPI_{Cl} and *CWV* indices.

Whereas Fig. 2.21 indicates a degree of correlation between the crack distribution indices and the total chloride content, the IPI_{Cl} and *CWV* indices have not been derived or weighted for chloride ingress into non-saturated, cracked SHCC, as occurs in cyclic wetting and drying by means of ponding or capillary absorption of NaCl solution. For instance, the IPI_{Cl} incorporates a threshold crack width of 30 μm for steady-state chloride diffusion through water-filled parallel cracks (Boshoff et al. 2016; Djerbi et al. 2008). Below the threshold mentioned, no chloride passes through the cracks, while the steady-state diffusion is considered to be proportional to the crack width for crack widths between 30 and 100 μm, beyond which it remains constant. Further research is required to derive an ingress index that captures the mechanism of chloride ingress into flexural cracks in non-saturated SHCC more directly than above. Nevertheless, reasonable correlation is found

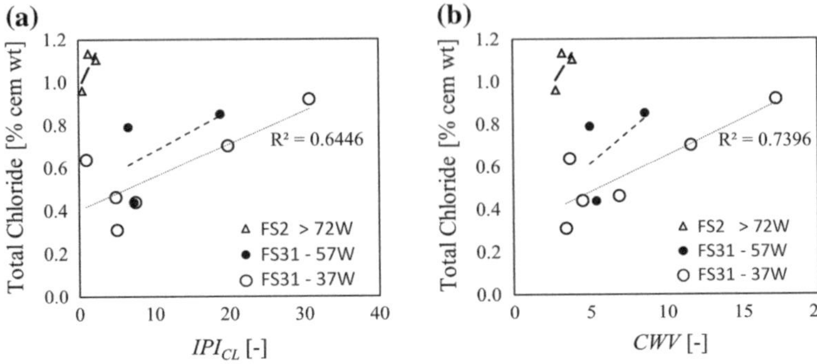

Fig. 2.21 Total chloride content (XRF) at the steel centre as function of **a** Ingress Potential Index (IPI_{Cl}), and **b** crack width value (CWV)

between the chloride content at the level of the steel bar in cracked SHCC and both the IPI_{Cl} and the CWV, for shorter exposure (37 weeks). In Chap. 9, test results of chloride-induced corrosion of rebar in the specimens (R/FS2, R/FS31) used are discussed, and the correlations between actual corrosion damage and the indices mentioned are investigated.

2.6 Capillary Absorption

Capillary absorption appears to be a frequently used criterion for the durability of cement-based materials. A high capillary absorption results in an accelerated ingress of deleterious substances. Under cyclic exposure, the substances can accumulate, over time, within the material. This is of particular importance for the ingress of chlorides.

Capillary absorption may be experimentally investigated by means of standardised test methods, such as according to DIN EN ISO 15148 (2003). It is usually assumed that the water uptake is proportional to the square root of the time t, and, accordingly, a so-called water absorption coefficient A is defined as follows:

$$A = \frac{x(t)}{\sqrt{t}} \cdot u_f \cdot \rho_{Solid} \tag{2.16}$$

with:

$x(t)$ the time-dependent penetration depth
u_f the maximum water content
ρ_{Solid} the dry density of the material

The capillary absorption of SHCC has been the subject of multiple investigations. Mechtcherine et al. (2007) reported on tests of undamaged and cracked

SHCC. The water absorption coefficient after 24 h was related to the total crack length, enabling a proportionality to be found. On the basis of the experimental results, deviations from the above-mentioned and generally accepted $\sqrt{t}$ dependence may be observed. Functions containing the fourth root of time allow for the improvement of fits to measured data, especially when long exposure times are considered. For ordinary concrete, deviations from the $\sqrt{t}$ dependence have been observed before.

In addition, Şahmaran and Li (2009) observed an increasing amount of water absorption, with an increasing damage level. The measured data also deviate from the above-mentioned $\sqrt{t}$ dependence. A linear relationship between water uptake and number of cracks could be shown. The application of a water-repellent admixture led to a significant reduction of the water absorption involved.

Zhang et al. (2010) were able to show, by means of the use of neutron radiography, that existing cracks, including those with small widths, at an SHCC surface subjected to capillary absorption become water-filled after only a few minutes and that, thereafter, the water penetrates from the cracks into the adjacent material. According to Wittmann et al. (2011), the effect is significantly reduced when the mortar matrix is hydrophobised by means of the use of a silane emulsion.

Similar observations were reported by Schröfl et al. (2015), who investigated the capillary absorption of RC beams with SHCC overlay (see Sect. 8.6). In addition to the fast water uptake of the cracks, the researchers concerned found an intensified capillary absorption for wider cracks.

The direct influence of crack distances on capillary absorption has not been the subject of the above-mentioned experimental observations. The influence concerned is of particular importance in the case of non-uniform crack spacing. Wagner (2016) conducted a test series aimed at separating individual contributions to capillary absorption, namely at detecting the capillary absorption of the undamaged SHCC surface, the water uptake of the cracks, and the capillary absorption of the rough crack surfaces. The crack spacing involved was believed to directly affect the overall absorption behaviour, since relatively narrow spaces between neighbouring cracks are likely to become saturated by water before wider ones are. In order to prove the assumption made, the tests that are schematically shown in Fig. 2.22 were conducted.

Below, the results obtained by Wagner (2016) are summarised. The absorbed water mass of the undamaged material appeared to be proportional to the fourth root of the exposure time. No essential difference was found between the mortars with and without fibres (see Fig. 2.23).

Figure 2.24a shows the expected linear relationship between the absorption coefficient after 24 h and the crack density, meaning the number of cracks per unit length. In Fig. 2.24b, it can be seen that the slope of the above-mentioned relationship decreases over the exposure time concerned. The decrease is attributed to the fact that the portion of the fully saturated spaces between neighbouring cracks increases.

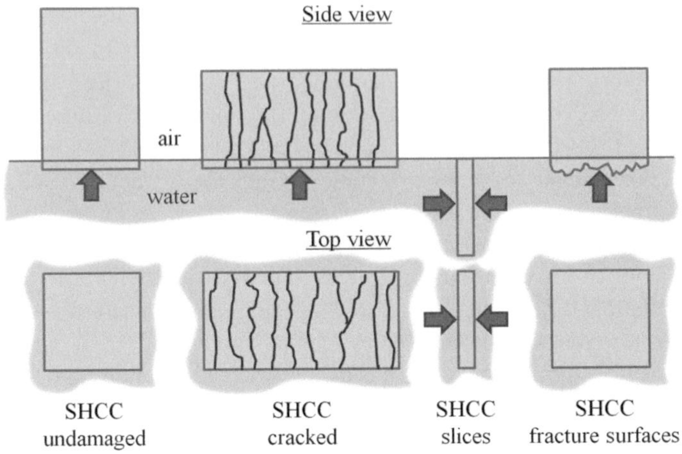

Fig. 2.22 Exposure conditions for capillary absorption tests (Wagner 2016)

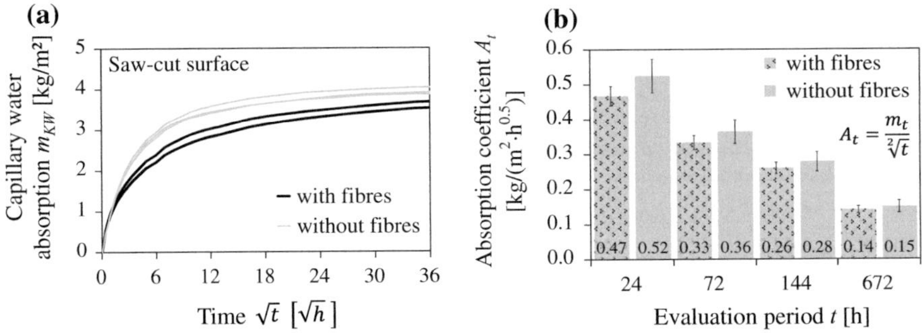

Fig. 2.23 **a** Water absorption of uncracked mortar versus time, and **b** absorption coefficient for different evaluation periods (Wagner 2016)

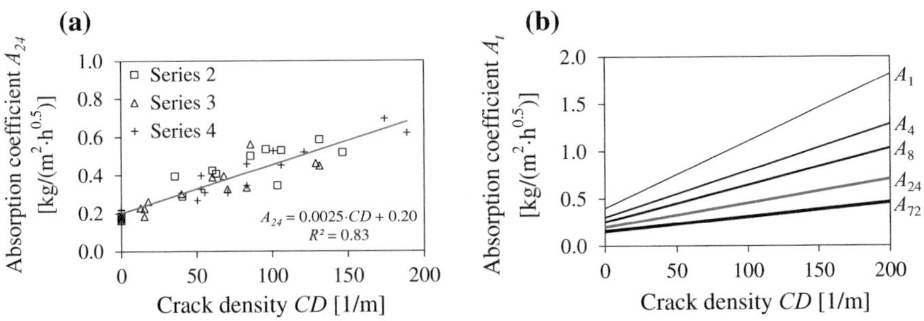

Fig. 2.24 Water absorption coefficient versus crack density for **a** an evaluation period of 24 h, and **b** for different evaluation periods (Wagner 2016)

The results for the capillary absorption of thin uncracked SHCC slices (see Fig. 2.22, third column) are presented in Fig. 2.25. At some point in time, depending on the slice thickness, it can clearly be seen that the water absorption is suddenly reduced. The reduction is due to the event reaching a state of water saturation.

The capillary absorption of fracture surfaces has also been investigated. In the tests concerned, the damage or micro-fracturing level of the surfaces was varied by means of using different fibre contents V_F. The higher the fibre content, the rougher or more micro-cracked the fracture surface is likely to be on the macroscopic scale. In Fig. 2.26a, it can be seen that the capillary absorption will be stronger for the rougher surfaces. The reasons for such an effect are the larger exposed surface area involved and, possibly, also the increased porosity due to micro-cracking. The diagram in Fig. 2.26b shows the capillary absorption coefficient A^{IV} for the case of proportionality to the fourth root of time. The influence of the fibre content, which is here assigned to a damage level D_F (see upper axis in Fig. 2.26b) can clearly be seen. The factor β_D, displayed in the same diagram, describes the relative increase of the capillary absorption coefficient with respect to the case of the fibreless mortar.

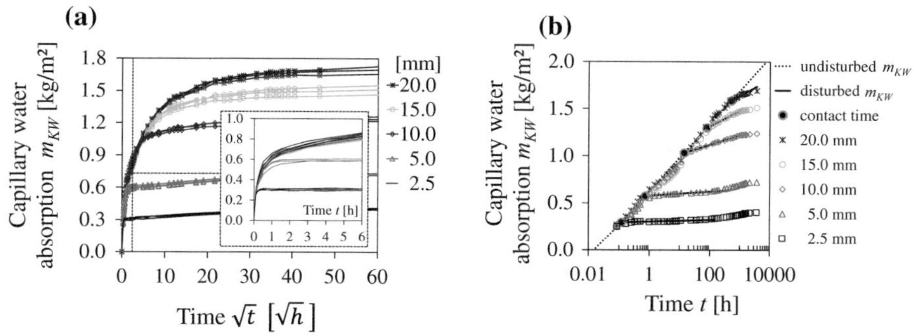

Fig. 2.25 Water absorption of thin SHCC slices (Wagner 2016)

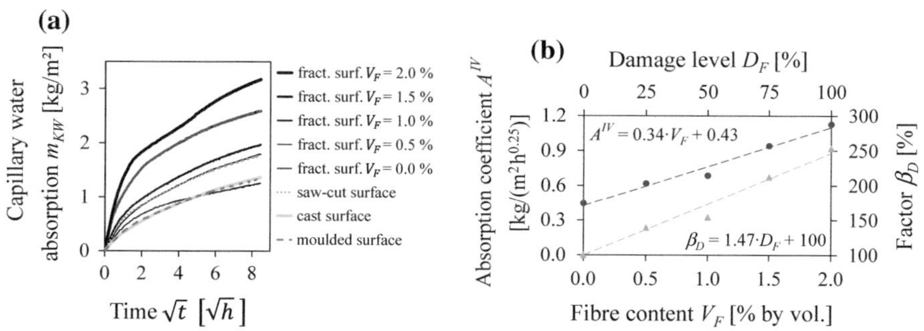

Fig. 2.26 Capillary absorption of fracture and reference surfaces (Wagner 2016)

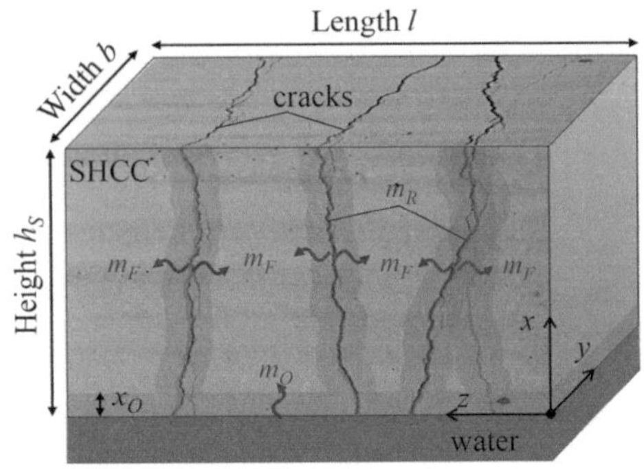

Fig. 2.27 Contributions to the capillary absorption of cracked SHCC (Wagner 2016)

The experimental observations formed the basis for the development of an analysis model in respect of the capillary absorption of porous materials with multiple closely spaced cracks (Wagner 2016). The above-mentioned three effects are schematically shown in Fig. 2.27. The different values m are always the absorbed mass per unit base area (at the lower surface).

The water uptake of the crack is denoted with m_R . By using the crack pattern parameters, as introduced in Sect. 2.2, including the class widths Δl and Δw, the following equation is obtained:

$$m_R = \frac{\rho_{Fluid} \cdot (\Delta l \cdot CLV) \cdot (\Delta w \cdot CWV) \cdot CD \cdot l \cdot h_S}{l \cdot b} \tag{2.17}$$

Note that m_R is independent of the exposure time involved. The independence is justified by the above-mentioned observation that the cracks are almost immediately water-filled after exposure.

For the capillary absorption of the fracture surfaces m_F, the following equation has been derived (Wagner 2016):

$$m_F(t) = \frac{\bar{A}_D^{IV} \cdot \overline{\beta_h}}{l \cdot b} \cdot \int_0^t \frac{1}{4} \cdot \tilde{t}^{-\frac{3}{4}} \cdot \beta_s(\tilde{t}) \cdot \overline{\beta_Q}(\tilde{t}) \cdot A_F(\tilde{t}) d\tilde{t} \tag{2.18}$$

with:

$$\mathrm{A}_D^{IV} = \mathrm{A}^{IV} \cdot \beta_D \text{ and } \beta_s(\tilde{t}) = \left(\frac{1}{1 - CSV \cdot \Delta s} + 1\right)^{\left(\frac{\left(\frac{2}{1-CSV \cdot \Delta s}+2\right) \cdot \bar{A}_D^{IV}}{u_f \cdot \rho_{trocken}} \cdot \sqrt[4]{\tilde{t}}\right)} \tag{2.19}$$

with:

A_D^{IV} the capillary absorption coefficient of the bulk material ("IV" for fourth root of time)

β_D the experimentally determined factor for the damage level of the fracture surfaces (see Fig. 2.26b)

$\beta_s(\tilde{t})$ a factor considering the experimentally determined crack spacing distribution CSV for a class width Δs

$\overline{\beta_Q}(\tilde{t})$ a factor considering the flow rate limited by the crack width

$\overline{\beta_H}$ a factor considering the changing pressure head along the crack

$A_F(\tilde{t})$ the crack surface area corrected for the penetration depth x_O at the lower surface (see Fig. 2.27)

The capillary absorption at the external (lower) surface may be calculated as follows:

$$m_O(t) = A^{IV} \cdot \int_0^t \frac{1}{4} \cdot \tilde{t}^{-\frac{3}{4}} \cdot \beta_s(\tilde{t}) \cdot d\tilde{t} \tag{2.20}$$

Finally, the theoretical total capillary absorption $m_{KW}^{th}(t)$ may be composed by means of applying the following equation:

$$m_{KW}^{th}(t) = m_O(t) + m_F(t) + m_R(t) \tag{2.21}$$

Similarly to the case of water permeability, the difference between the theoretical capillary absorption $m_{KW}^{th}(t)$ and the experimentally determined capillary absorption $m_{KW}^{exp}(t)$ may be quantified by means of the coefficient ξ_{KW}.

$$\xi_{KW} = \frac{m_{KW}^{exp}(t)}{m_{KW}^{th}(t)} \tag{2.22}$$

The reasons for the difference between theoretical and measured capillary absorption are the changing crack pattern parameters over the specimen depth, the existence of a nonconstant damage level along the crack length, and other deviations from the model simplifications. Figure 2.28 shows both the calculated and the

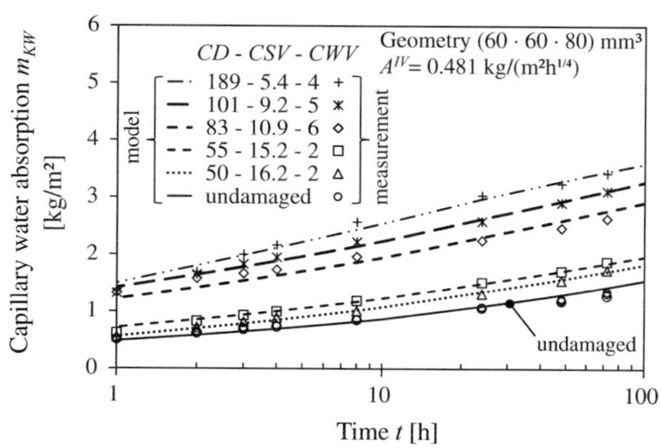

Fig. 2.28 Comparison of measured and calculated capillary water absorption of uncracked and cracked SHCC (Wagner 2016)

measured capillary absorption versus the exposure time. Fitting the calculated capillary absorption curves to the measured data yielded a coefficient $\xi_{KW} = 0.338$, with a standard deviation of 0.065.

2.7 Recommendations for Durability Design and Conclusions

The transport of different media through SHCC has been the subject of several experimental investigations. The relevant physical processes and the respective specifics of SHCC are known. In general, the crack patterns formed in the material appear to have a predominant influence on the transport rates and, if relevant, also on their dependence on exposure time. Hence, the combined action of mechanical loading and exposure to penetrating media has to be considered.

Methods for predicting the transport rates or for estimating the vulnerability to specific transport processes have been proposed. Some of the methods are empirical, others are based on known physical models, but all of them consider the influence of the actual crack pattern. Parameters were defined for describing and comparing the SHCC crack patterns. Such parameters may be determined on the basis of images of the respective cracked SHCC surfaces. In addition to the material properties, the type of the mechanical loading is affecting the formation of the crack pattern.

As far as chloride ingress is concerned, the proposed crack pattern parameters appear to be good indicators for shorter term ingress, and possibly even for corrosion initiation. But the accumulation of chloride in flexural cracks over time has not yet been correlated to these indices. This is partially due to limited experimental data being available and may provide further research opportunities in future. So far, no crack pattern parameter has yet been proposed that considers the crack depths as being of particular significance in relation to flexural loading.

The prediction of transport rates for certain exposure and loading conditions may form part of the durability design. An open issue, however, appears to be the quantification of the relationship between transport processes and structural durability. At the very least, critical values for the transport rates should be identified.

References

Boshoff, W.P., Adendorff, C.J. (2010). Modelling SHCC cracking for durability. V. Mechtcherine, M. Kaliske (eds.), Proceedings of a Symposium organized within the European Conference on Fracture (ECF-18), Aug. 30-Sept. 3, 2010, Dresden, Germany, Aedificatio Publishers, Freiburg, Germany, pp. 195-202.

Boshoff, W.P., Altmann, F., Adendorff, C.J., Mechtcherine, V. (2016). A new approach for modelling the ingress of deleterious materials in cracked strain hardening cement-based composites. Materials and Structures 49(6):2285-2295.

Clear, C.A. (1985). The effects of autogenous healing upon the leakage of water through cracks in concrete. Technical Report 559, Cement and Concrete Association.

DIN EN ISO 15148:2003-03 (2003). Wärme- und feuchtetechnisches Verhalten von Baustoffen und Bauprodukten – Bestimmung des Wasseraufnahmekoeffizienten bei teilweisem Eintauchen. Deutsches Institut für Normung, March 2003.

Djerbi, A., Bonnet, S., Khelidj, A., Baroghel-Bouny, V. (2008). Influence of traversing crack on chloride diffusion into concrete. Cement and Concrete Research 38(6):877-883.

DuraCrete (2000). Probabilistic performance based durability design of concrete structures. Report R15, 2000, Project BE95-1347, The European Union – Brite Euram III.

Edvardsen, C.K. (1996). Wasserdurchlässigkeit und Selbstheilung von Trennrissen in Beton. Deutscher Ausschuss für Stahlbeton, Heft 455, Beuth Verlag, Berlin, Germany.

Kobayashi, K., Iizuka, T., Kurachi, H., Rokugo, K. (2010). Corrosion protection performance of high performance fibre reinforced cement composites as a repair material. Cement and Concrete Composites 32(6):411-420.

Kojima, Y., Dung, L.A., Rokugo, K., Kobayashi, K. (2014). Chloride proofing and rebar corrosion proofing performances of SHCC having cracks. E. Schlangen, M.G. Sierra Beltran, M. Luković, G. Ye (eds.), Proceedings of the 3rd International RILEM Conference on Strain Hardening Cementitious Composites, 3-5 November 2014, Dordrecht, The Netherlands, RILEM Publications S.A.R.L., Bagneux, France, pp. 61-68.

Lepech, M.D., Li, V.C. (2009). Water permeability of engineered cementitious composites. Cement and Concrete Composites 31(10):744-753.

Mechtcherine, V., Lieboldt, M., Altmann, F. (2007). Preliminary tests on air-permeability and water absorption of cracked and uncracked strain hardening cement-based composites. K. Audenaert, L. Marsavina, G. de Schutter (eds.), Proceedings of the International RILEM Workshop on Transport Mechanisms in Cracked Concrete, Ghent, Belgium, 7 September 2007, Acco, Leuven, Belgium, pp. 55-66.

Mechtcherine, V., Lieboldt, M. (2011). Permeation of water and gases through cracked textile reinforced concrete. Cement and Concrete Composites 33(7):725-734.

Moukwa, M (1990). Deterioration of concrete in cold sea waters. Cement and Concrete Research 20(3):439-446.

Paul, S.C., Van Zijl, G.P.A.G. (2014). Crack formation and chloride induced corrosion in reinforced strain hardening cement-based composite (R/SHCC). Journal of Advanced Concrete Technology 12:340-351.

Paul, S.C. (2015). The role of cracks and chloride in corrosion of reinforced strain hardening cement-based composite (R/SHCC). PhD Thesis, Stellenbosch University, South Africa.

Paul, S.C., Van Zijl, G.P.A.G., Babafemi, A.J., Tan, M.J. (2016). Chloride ingress in cracked and uncracked SHCC under cyclic wetting-drying exposure. Construction and Building Materials 114:232-240.

RILEM TC 178-TMC (2002a). Testing and modelling chloride penetration in concrete, Analysis of total chloride content in concrete. Materials and Structures 35(9):583-585.

RILEM TC 178-TMC (2002b). Testing and modelling chloride penetration in concrete, Analysis of water soluble chloride content in concrete. Materials and Structures 35(9):586-588.

Ripphausen, B. (1989). Untersuchung der Wasserdurchlässigkeit und Sanierung von Stahlbetonbauteilen mit Trennrissen. Doctoral thesis, RWTH Aachen, Germany.

Şahmaran, M., Li, V.C. (2009). Influence of microcracking on water absorption and sorptivity of ECC. Materials and Structures 42(5):593-603.

Schröfl, C., Mechtcherine, V., Kaestner, A., Vontobel, P., Hovind, J., Lehmann, E. (2015). Transport of water through strain-hardening cement-based composite (SHCC) applied on top of cracked reinforced concrete slabs with and without hydrophobization of cracks – Investigation by neutron radiography. Construction and Building Materials 76:70-86.

Tsukamoto, M., Fiebrich, M (1994). Untersuchung zur Durchlässigkeit von faserfreien und faserverstärkten Betonbauteilen mit Trennrissen. Deutscher Ausschuss für Stahlbeton, Heft 440, Beuth Verlag, Berlin, Germany.

Van Zijl, G.P.A.G., Slowik, V., Toledo Filho, R.D., Wittmann, F.H., Mihashi, H. (2016). Comparative testing of crack formation in strain-hardening cement-based composites (SHCC). Materials and Structures 49(4):1175-1189.

Wagner, C., Bretschneider, N., Slowik, V. (2011). Evaluation of crack patterns in strain hardening cement-based composites (SHCC) with respect to structural durability. Restoration of Buildings and Monuments 17(3/4):223-237.

Wagner, C., Slowik, V. (2011). On the water permeability of cracked strain hardening cement-based composites. R.D. Toledo Filho, F.A. Silva, E.A.B. Koenders, E.M.R. Fairbairn (eds.), Proceedings of the 2nd International RILEM Conference on Strain Hardening Cementitious Composites (SHCC2), 12-14 December 2011, Rio de Janeiro, Brazil, RILEM Proceedings (PRO 81), RILEM Publications S.A.R.L., Bagneux, France, pp. 181-188.

Wagner, C., Dollase, A., Slowik, V. (2012). Evaluation of crack patterns in SHCC with respect to water permeability and capillary suction. M.G. Alexander (ed.), Proceedings of the 3rd International Conference on Concrete Repair, Rehabilitation and Retrofitting (ICCRRR), 3-5 September 2012, Cape Town, South Africa, CRC Press, Leiden, The Netherlands, pp. 972-977.

Wagner, C. (2016). Dauerhaftigkeitsrelevante Eigenschaften von dehnungsverfestigenden zementgebundenen Reparaturschichten auf gerissenen Betonuntergründen (Durability related properties of strain hardening cement-based repair layers on cracked concrete substrates). Doctoral thesis, Technische Universität Dresden, Germany.

Wang, K., Jansen, D.C., Shah, S.P., Karr, A.F. (1997). Permeability study of cracked concrete. Cement and Concrete Research 27(3):381-393.

Wittmann, F.H., Wang, P., Zhang, P., Zhao, T., Beltzung, F. (2011). Capillary absorption and chloride penetration into neat and water repellent SHCC under imposed strain. R.D. Toledo Filho, F.A. Silva, E.A.B. Koenders, E.M.R. Fairbairn (eds.), Proceedings of the 2nd International RILEM Conference on Strain Hardening Cementitious Composites (SHCC2-Rio), 12-14 December 2011, Rio de Janeiro, Brazil, RILEM Proceedings (PRO 81), RILEM Publications S.A.R.L., Bagneux, France, pp. 165-172.

Wittmann, F.H., Zhang, P., Zhao, T. (2012). Importance of water repellent treatment of SHCC for durability and service life. Restoration of Buildings and Monuments 18(1):31-40.

Yeomans, S.R. (2004). Galvanized steel reinforcement in concrete. Elsevier Science, Amsterdam, The Netherlands.

Zhang, P., Wittmann, F.H., Zhao, T., Lehmann, E.H. (2010). Neutron imaging of water penetration into cracked steel reinforced concrete. Physica B: Condensed Matter 405 (7):1866-1871.

Chapter 3
Fiber Durability

Flavio A. Silva, Alva Peled, Bartosz Zukowski and Romildo D. Toledo Filho

Abstract The service life of strain-hardening cement-based composite materials (SHCC) is based on the service life of all the system components: fiber, matrix, and fiber–matrix interface. This chapter describes SHCC durability from the fiber perspective, distinguishing between different fiber types commonly used in SHCC: polyvinyl alcohol (PVA), polyethylene (PE), polypropylene (PP), and natural, steel and glass. Their relative strengths should be considered during any design process, especially when determining the long-term durability of the composite material. PVA is considered a chemically stable fiber, resistant to acid solutions, organic solvents, and alkaline environments. It has proven long-term durability, based on its high strength retention after accelerated aging cycles. The fibers are also hydrophilic as the cement matrix, leading to strong bonding between the two. PE and PP, both olefin type fibers, are also known for their high durability performance in the cement matrix. However, these fibers are hydrophobic and therefore do not have any chemical affinity to the hydrophilic cement matrix, resulting in low bonding between these fibers and the cement matrix. Glass fibers are relatively sensitive to the alkaline environment of the cement matrix. Consequently, special glass fibers (known as AR or alkali-resistant fibers) have been developed with improved alkali resistance. Steel fibers have a good affinity with the cement matrix but, depending on the environmental conditions, can undergo corrosion. The new trend of natural fibers in SHCC design offers a major advantage: combining low cost with local availability. However, they are susceptible to weathering, alkaline environments, biological attack, and mineralization.

Keywords Fiber · Durability · Thermal effects · Chemical effects

---

F.A. Silva (✉)
Pontifícia Universidade Católica do Rio de Janeiro, Rio de Janeiro, Brazil
e-mail: fsilva@puc-rio.br

A. Peled
Ben-Gurion University of the Negev, Beersheba, Israel

B. Zukowski · R.D. Toledo Filho
Universidade Federal do Rio de Janeiro, Rio de Janeiro, Brazil

G.P.A.G. van Zijl and V. Slowik (eds.), *A Framework for Durability Design with Strain-Hardening Cement-Based Composites (SHCC)*, RILEM State-of-the-Art Reports 22,
DOI 10.1007/978-94-024-1013-6_3

3.1 Introduction

Strain-hardening cement-based composite (SHCC) is a material designed to achieve strain-hardening properties by mixing short, high-strength fibers into a fine mortar, with a maximum aggregate size of typically 0.3 mm. The methods of achieving strain-hardening properties suggest the use of different types of fibers, among which are polyvinyl alcohol (PVA), polyethylene (PE), polypropylene (PP), glass, and natural or steel fibers (see Table 3.1). The mixture proportioning is based on the fiber properties (strength and interfacial properties) combined with the matrix fracture energy. The fiber type used in the design could be based on its availability, on its local market-related cost (i.e. the prevailing economic conditions), and on the weathering exposition of the produced composite material (for instance, its freeze/thaw cycles, its location in humid areas, or its exposure to a high temperature gradient). The durability of the fibers used as reinforcement in SHCC could be regarded from one of two perspectives. The first relates to the durability of the fiber in the matrix environment (alkaline), whereas the second relates to the durability of the fiber that is exposed to different environmental conditions (i.e. high temperatures and humidity). This chapter focuses on the durability of various types of fibers that may be used in SHCC.

Table 3.1 Mechanical properties of commonly used fibers in SHCC[a]

Type of fiber	Equivalent diameter (mm)	Specific gravity (kg/m^3)	Tensile strength (MPa)	Young's modulus (GPa)	Ultimate elongation (%)
Polyvinyl alcohol (PVA)	0.027–0.66	1300	900–1600	23–40	7–8
Polyethylene (PE)	0.025–1.0	960	200–300	5.0	3.0
Polypropylene (PP)	0.02–0.40	950	550–760	3.5	15–25
Sisal	0.1–0.2	900	286–526	19	2.6–5.2
Jute	0.1–0.2	1030	160–440	35–48	0.6–0.9
Glass	0.005–0.15	2500	1000–2600	70–80	1.5–3.5
Steel	0.15–1.00	7840	345–3000	200	4–10

[a]Data from Balaguru and Shah (1992), Fidelis et al. (2013), Kuraray (2015), Nawy (1996), Saechtling (1987), Silva et al. (2008)

Notes 1 mm = 0.039 in.; 1 kg/m^3 = 0.06 lb/ft^3; 1 MPa = 145 psi; 1 GPa = 1,450,000 psi

3.2 Durability of PVA Fiber

The PVA fiber belongs to the category of high Young's modulus fibers, in which the division between high and low is based on the modulus of the matrix concerned (Bentur and Mindess 2007). Several types of PVA fibers can be found in the market (see Table 3.2), among which is the REC15, which was especially designed for use in SHCC. Due to the variety of PVA fiber types, there is variation in its tensile strength, in Young's modulus, and in its diametrical dimensions. However, most PVA fibers are characterized by the existence of a high degree of bonding with the matrix, and with resistance to an alkaline environment. The two characteristics mentioned suggest that SHCC that is produced using PVA fibers should be extremely durable. Hoshiro et al. (2006) correlated the strength retention of PVA fiber in hot alkaline water to tensile strength variation and ultraviolet (UV) absorption. Based on their calculations, the estimated tensile strength of such fiber after 60, 100 and 120 years is 100, 99.8 and 97.5%, which suggests that PVA fibers should have long-term durability. Moreover, PVA fiber is considered chemically stable, showing high strength retention in acid solutions and organic solvents. Several durability tests have already been carried out on PVA fibers (De Lhoneux et al. 2002; Li et al. 2004), especially on the REC15 type. The results have shown that the durability of the composite is related to the change in the interfacial parameters present due to weathering.

The PVA fiber also shows strength retention after exposure to temperatures up to 150 °C (Magalhães et al. 2010, 2015). However, the SHCC that was reinforced

Table 3.2 Properties of Kuralon PVA fiber

Types	Product	Diameter (mm)	Tensile strength (MPa)	Young's modulus (GPa)	Elongation capacity (%)	Main application
Standard	RM182	0.014	1500	36	7	Cement board, mortar
	RSC15	0.04	1400	36	6.5	Mortar, concrete
	RF400	0.20	975	27	9	Mortar, concrete
	RF1000	0.31	975	26	6	Mortar, concrete
	RF4000	0.67	900	23	9	Concrete
Ductile	RECS7	0.27	1560	39	6.5	Cement board, mortar
	REC15	0.04	1600	41	6.5	Mortar
	RECS100	0.10	1200	28	12.5	Mortar, concrete

with PVA fibers that were submitted to moderate temperatures under tensile loading showed a decline in mechanical behavior (Oliveira et al. 2014).

3.3 Durability of PE and PP Fiber

PE fiber that is used as a net/textile or fibrils can increase the load-bearing capacity of the composite material. Due to the weak bond to the cement matrix, PE fibers require treatment to be applied into the composite. One technique of increasing the bonding entails their deformation by means of fibrillation, crimping, or twisting. Another technique involves plasma treatment of the fiber surface that can enhance the bond strength by 100% in comparison with that of untreated fiber. Both techniques can be used in combination to improve the results obtained and to increase the durability of the composite material involved (Brandt et al. 1994).

PP fibers are highly resistant both to most chemicals and to the high alkaline environment of the cement matrix, which makes them attractive for use in cement-based composites (Bentur 1989; Hannant and Zonsveld 1980).

3.4 Durability of Glass Fiber

Several types of glass fiber, differing in their composition, have been developed and are available, including E-glass (with high electrical resistivity), S-glass (with high strength), and C-glass (with chemical resistance). However, the glass fibers mentioned are sensitive to the alkaline environment that is typical of cement-based matrices (pH = 12.5–13), resulting from the presence of the hydration products of calcium hydroxide [$Ca(OH)_2$] within the hardened cement paste matrix. Under high pH conditions, glass fibers are sensitive to chemical attack, leading to the breakdown of the Si–O–Si network of the glass fiber, and, thus, to the formation of defects on the surface that, in turn, lead to reduction in fiber strength (Hull 1981). Such instability of the glass fibers led to the development of alkali-resistant (AR) glass fibers for use in Portland cement-based composites. The alkali resistance involved is achieved by adding 15% zirconia (ZrO_2), by mass, to the glass fiber. The basic mechanical properties of AR glass and of E-glass fibers are provided in Table 3.3, with E-glass fibers being the most popular fibers for polymer-based composites.

Table 3.3 Basic mechanical properties of AR glass and E glass fibers (Gupta 1988; Majumdar and Nurse 1974)

	E	AR
Density (g/cm^3)	2.54–2.55	2.78
Tensile strength (MPa)	3500	2500
Modulus of elasticity (GPa)	72.4–76	70
Elongation at break (%)	4.8	3.6

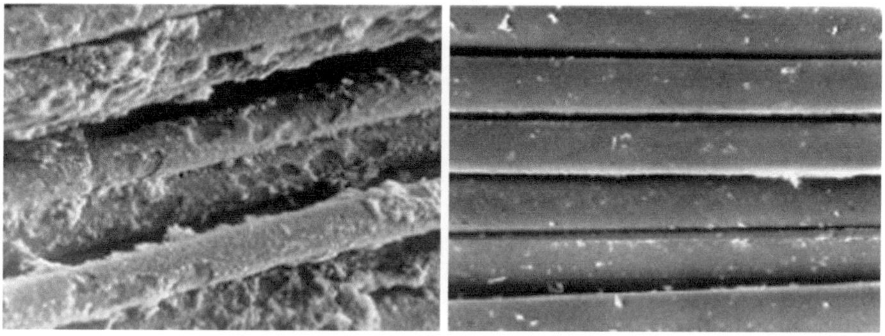

Fig. 3.1 Glass filaments exposed to an alkaline environment: **a** E-glass, and **b** AR glass (Masthoff 1998)

Figure 3.1 clearly presents the sensitivity of E-glass fibers to the alkaline environment of a Portland cement matrix, comparing the surfaces of E-glass fibers and AR glass fibers after exposure to alkaline conditions. Table 3.4 shows the mechanical performance of cement-based composites that have been reinforced with the two glass fibers concerned. E-glass is observed as suffering from aggressive corrosion in the alkali solution, presenting a rough and damaged surface, as compared to the smoother, cleaner, and more perfect surface of the AR glass. The surface damage and corrosion cause a significant decrease, during aging, in the mechanical properties of the cement-based composite reinforced with E-glass fibers compared to in the case of the AR glass fibers (Table 3.4) (Masthoff 1998). Interestingly, although, initially, the composite with E-glass can be seen to perform better, the stable behavior during the aging of the AR glass composite shows that the latter's overall performance in cement-based composites is superior to that of the former. However, due to the good bonding of AR glass fibers with cement-based matrices and their relatively low cost, which provides a good cost–performance ratio, AR glass fibers are used relatively commonly as reinforcing material in cement-based composite applications.

Glass fibers are available in short or continuous form, of which the latter can appear either as individual yarns or as assembled into technical fabrics. Continuous glass yarns are manufactured as multifilament bundles (i.e. roving) that consists of hundreds or thousands of filaments. Such bundle form is problematic when it is

Table 3.4 Tensile strength (MPa) of E and AR glass fibers in concrete during accelerated aging of the test samples (stored in hot water at 80 °C) (Masthoff 1998)

Reinforcement	Age of the test samples (in hours) stored at 80 °C								
	0	25	26	27.5	30.5	34	48	72	96
AR glass	840	846	753	742	810	820	853	793	778
E-glass	1250	1072	926	878	739	608	362	274	197

used as reinforcement for cement-based matrices, as the cement matrix concerned consists of relatively large particles (~ 10 μm) that are larger than the spaces between the filaments of the bundle, preventing the cement particles from being able fully to penetrate the inter-bundle spacing. The result of the above is unique bonding mechanisms, whereby the external filaments ('sleeve') are in intimate contact with the hydration products, whereas the internal filaments ('core') are relatively free (Banholzer 2004; Bentur and Mindess 2007; Brameshuber et al. 2006; Zhu and Bartos 1997) (Fig. 3.2a). Due to the special microstructure of the reinforcing bundle, the external filaments (i.e. the sleeve filaments), which are in direct contact with the cement matrix and which are well-bonded to the matrix, are fractured during loading, providing high levels of first-crack stress. After their failure, a telescopic type of pull-out is generated, whereby the internal filaments (i.e. the 'core' filaments) slip against the external ('sleeve') filaments (Fig. 3.2b).

The sliding mode of the inner filaments induces high ductility in the composite that is favorable for brittle glass-fiber-reinforced cement-based composite (GFRC) systems. However, during continued hydration and aging, a slow and gradual process of deposition of hydration products between the filaments takes place, which can change the nature of bonding by means of increasing the sleeve/core ratio. The change results in the strengthening of the bond, which, although being a favorable effect, can also lead to embrittlement, when brittle glass fibers are used (Bentur and Mindess 2007; Zhu and Bartos 1997; Bentur and Diamond 1987). The

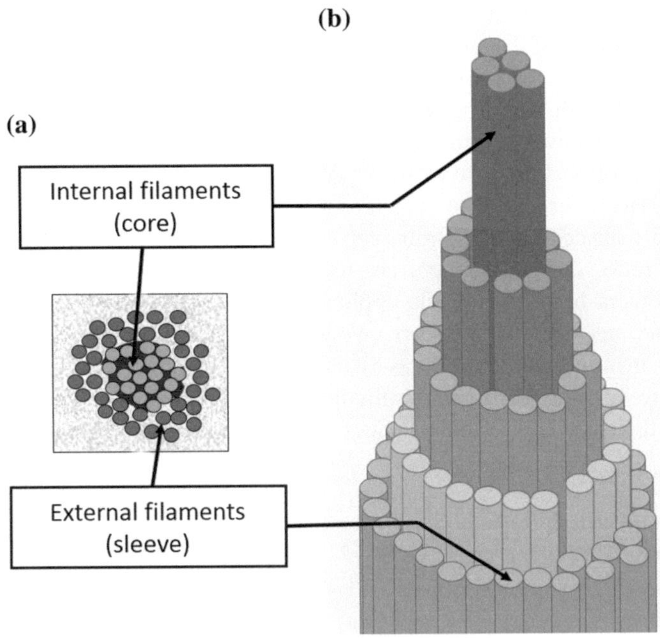

Fig. 3.2 **a** Schematic cross-section of a multifilament bundle embedded in a cement matrix; **b** schematic view of the telescopic pull-out under tensile loads

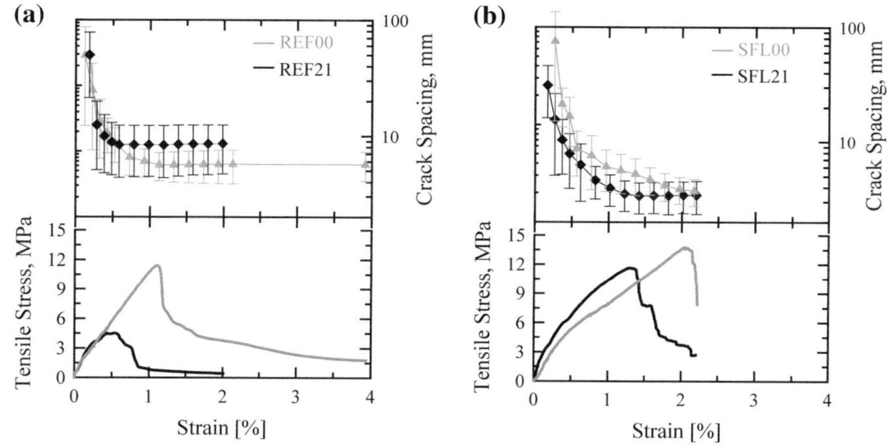

Fig. 3.3 Effect of aging on tensile properties and crack spacing of cement composites reinforced with AR glass fabrics: **a** a reference composite (REF); and **b** a composite with micro-sized silica fume coating (SFL), before aging (00) and after aging (21) at 50 °C for 21 days (Cohen and Peled 2010)

above means that, as the bonding changes over time, so too do the composite properties and durability involved. Another change that takes place in GFRC over time is caused by the chemical attack of the glass fibers, when they are surrounded by cement hydrates (Raupach et al. 2006a). The diffusion process of the cement products over time, when occurring between the filaments of the bundle, can lead to the intensive corrosion of the glass filaments concerned. Since an increased number of fibers are in direct contact with the hydrate products of the cement paste matrix, the reduction in the overall behavior, toughness, and strength of the GFRC is more severe. To improve glass strand durability within the cement-based matrix, the roving requires coating. Many studies have been dedicated to the effects of coating on the glass bundles and on filling in the bundle gaps with polymers. A wide range of polymers, with varied viscosity properties, have been studied for their fabric-bundle-coating abilities in cementitious composite applications (Butler et al. 2009; Gao et al. 2004; Glowania et al. 2011; Raupach et al. 2006b).

Another type of coating that has also been investigated is made of silica fume particles (Bentur 1989; Bentur et al. 2010, 2013; Cohen and Peled 2010, 2012). In such cases, silica fume particles of nano and micro size were inserted into the bundle spaces by means of the impregnation process, improving the transfer taking place between the inner filaments and the cement matrix, so as to improve both the bonding and the overall performance of the composite. The silica fume particles were found to react with the cement hydration products (i.e. calcium hydroxide), with their pozzolanic reaction leading to greater bonding and improved tensile behavior over time (Fig. 3.3b) as compared to the same properties of cement composite in which the glass bundles are uncoated (Fig. 3.3a). Due to the presence of the silica fume particles in between the bundle filaments, the pozzolanic reaction deposition of the cement products between the bundle filaments was found to decrease over time, allowing for the retention of the desired properties of the composite, even past the late stages involved (Fig. 3.3b). Furthermore, the silica-fume-coated AR glass composite, after aging (Fig. 3.3b), exhibited greater strength and energy than did the unaged REF system (Fig. 3.3a). Such was reported mainly when micro-sized silica fume particles were used. Polymer-based coating was also found to improve the performance of cement-based composites reinforced with AR glass fabrics over time. However, in such instances aggressive delamination between the coated bundle and cement matrix was also reported, with the delamination limiting the practical use of the material concerned (Dvorkin 2014). Another important advantage of using silica fume, as opposed to a polymer-based coating, is the ability of the former to withstand high temperatures and fire.

3.5 Durability of Steel Fiber

The durability of steel fiber is correlated with the corrosion process caused by the chlorides or the pH reduction, due to the matrix carbonation. Due to the small cover for the fibers at the surface, some corrosion is likely to occur close to the surface, which can lead to mistaken conclusions about the composite material as a whole (Brandt et al. 1994). However, the multi-cracking mechanism can prevent further corrosion of the composite due to microcell, rather than macrocell, formation, and the existence of a large anode/cathode area ratio (JSCE 2008) (see Fig. 3.4).

The corrosion process for a single fiber can have much more deteriorative effect than in a steel bar, because of the fiber diameter involved (Kosa and Naaman 1990). Jin et al. (2001) indicated that bars with various diameters had the same thickness of corrosion layer when in the same corrosive condition. The conclusion drawn was

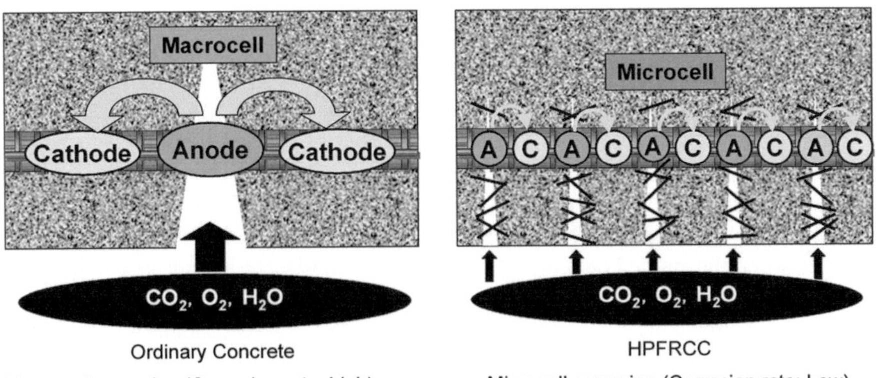

Fig. 3.4 Corrosion cell pattern differences in accordance with crack development mode (JSCE 2008)

that the fibers, because of the small diameter and because of a similar thickness of corrosion layer, should suffer much more from the corrosion than do the steel bars. For this reason, the protection of steel fibers should receive a great deal of attention. The fibers in concrete are protected by means of the dense structure (i.e. physical protection) and by means of the alkalinity of concrete (i.e. chemical protection) (Almusallam 2001). The penetration of chloride ions above a threshold value for corrosion of the steel fiber can destroy the protection otherwise accorded. Even in highly unfavorable environmental conditions (such as de-icing salt and seawater) only steel fibers with shallow cover and that are close to the surface, where the access of chloride ions is possible, are likely to be susceptible to corrosion (Balouch et al. 2010; Hoff 1987).

An important factor that influences the corrosion of fibers is the crack width, which increases the possibility of chloride penetration. The tests carried out by Granju and Balouch (2005) on the cracked steel-reinforced concrete (SFRC) in a marine-like environment for the time period of one year indicated the absence of corrosion in instances where the crack widths were less than 0.1 mm. The corrosion was progressive with the crack width. The specimens verified in tensile tests showed relatively high tensile strength, which was correlated with improved interference between the fiber and the matrix concerned, due to the roughness of the corroded layer involved. However, when the corrosion increased, the strength decreased.

The drying–wetting cycles can influence the corrosion of steel fibers, as can be seen in Fig. 3.5 (Waweru 2011). The progressive corrosion is easily noticed in fibers submitted to the aging cycles, in comparison with the reference fibers. After surface cleaning, the fibers exposed both many pits and a relatively high degree of corrosion (Fig. 3.6).

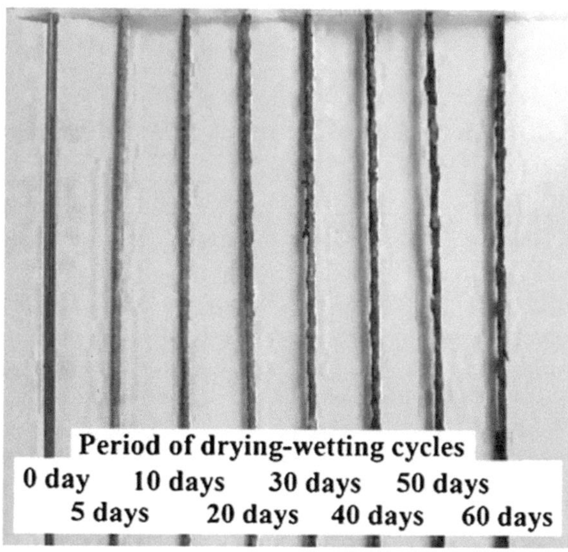

Fig. 3.5 Steel fibers after exposing to drying–wetting cycles (Waweru 2011)

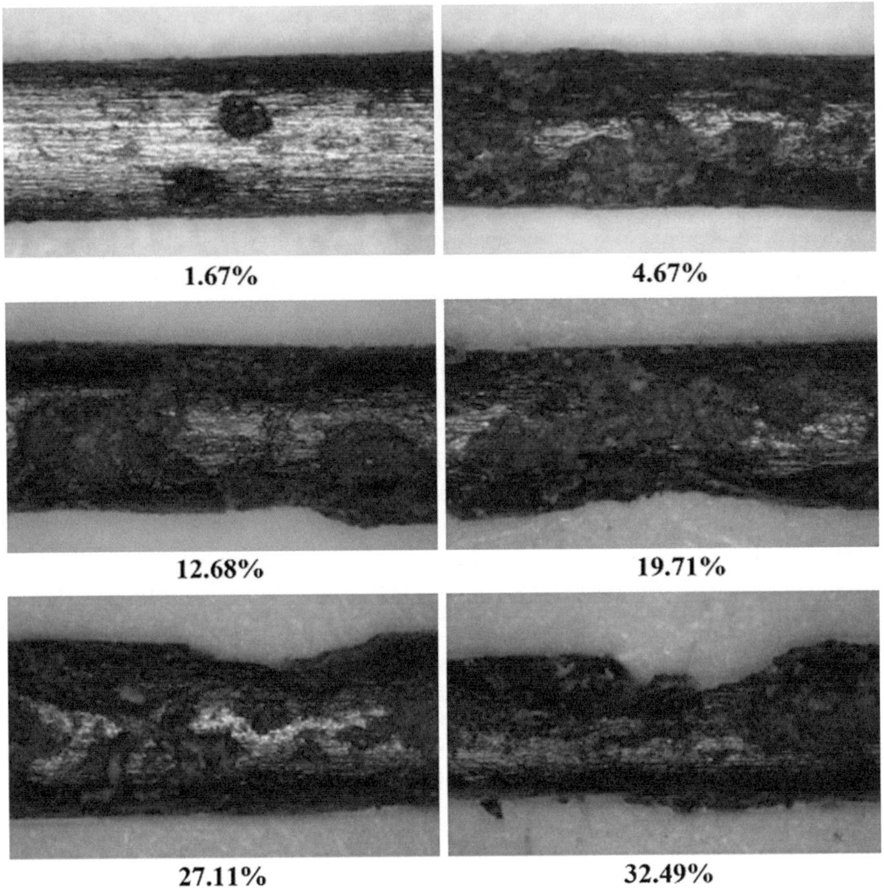

Fig. 3.6 Corrosion pitting of steel fibers with varying degrees of corrosion (Waweru 2011)

3.6 Durability of Natural Vegetable Fibers

Natural vegetable fibers are of interest in SHCC application, because of their local availability, their adequate mechanical performance, and their possible economic benefits (Fidelis et al. 2013; Melo Filho et al. 2013; Silva et al. 2008; Toledo Filho et al. 2000, 2003). Although they are much cheaper than any other fiber, unfortunately the properties of the fiber differ among the families of fibers, as well as within the same family, depending on the condition of the plant during growing processes (humidity, temperature, plant nutrition, insolation, or even the vicinity of other plants) (Oliveira et al. 2008). The possible application of natural fibers in SHCC is likely to be limited to the group of high-performance natural fibers (i.e. sisal, hemp and curauá) that present similar strength and similar Young's modulus to those of commonly used fibers (PVA and PE). Another possibility of application lies in combining different types of natural fibers at different scales. However, the research undertaken along such lines is still in its initial stage, with only a few research groups being, as yet, involved in related studies (PUC-Rio, COPPE/UFRJ and TU Delft). Other drawbacks of natural fiber application are related to the nature of the fibers concerned. Natural fibers are susceptible to natural weathering (i.e. to moisture movement), to degradation in an alkaline environment, to biological attack, and to mineralization in the matrix.

Natural fibers, which are mainly composed of cellulose, lignin and hemicellulose, also tend to contain pectin, oil and waxes (John and Thomas 2008; Wong et al. 2010). Each fiber is a combination of fibrils with layered structure (consisting of a primary cell wall and of three secondary walls around the lumen). The middle layer of the secondary cell walls determines the mechanical properties of fiber involved (being correlated with the microfibrillar angle and with the cellulosic content). The microfibrillar angle is the angle between the axis of the fiber and the microfibrils. The smaller the angle the higher the strength and stiffness are. The cellulose (which is a natural polymer) is the most important structural component that is found in natural fibers. Although the component is resistant to hydrolysis, alkaline and oxidizing agents, it is susceptible to chemical treatments. Hemicelluloses are lower molecular-weight polysaccharides and they play the role of a cementing material between cellulose microfibrils (Silva et al. 2008; Azwa et al. 2013). Lignin, which is a complex hydrocarbon polymer, is responsible for the rigidity of the plant and for water transport, as it has a hydrophobic nature. Lignin is also responsible for the yellowish color of the fiber, and, although it does not improve the strength of the fibers concerned, it is resistant to the onslaught of most microorganisms. It is soluble in alkali solutions and can be removed in the bleaching process. Wax and oils are present only on the fiber surface and function as a protection (Oliveira et al. 2008). They can be removed by means of hot water treatment, as well as by means of alkali solution immersion (Toledo Filho et al. 2000). The natural fibers with higher cellulose content, higher degree of polymerization, longer cell length and lower microfibrillar angle present higher mechanical performance than do other natural fibers (John and Thomas 2008; Silva et al. 2008). The application of natural fibers in cement composites is related to a relatively high risk of fiber degradation in comparison with the risk that is posed in the case of synthetic fibers (i.e. PVA). The degradation of the fiber concerned largely depends on the environmental conditions prevailing at the time.

3.6.1 Influence of Humidity

The moisture absorption of natural fibers largely depends on their hemicellulose content (Methacanon et al. 2010), with the higher the content of hemicelluloses, the higher the sorption of water involved (John and Thomas 2008). The degree of porosity that is related to the voids in natural fibers has an influence on the durability of the fibers concerned. The water movement can occur along the fiber, from the matrix to the fiber, as well as in the opposite direction. The water movement in the fiber causes swelling and shrinkage of the fiber, which can have an influence on the fiber–matrix bond involved.

In composites with low humidity, the natural fibers present a fragile failure mechanism, whereas, in saturated ones, the mechanism governing the damage is governed by fiber pull-out. Also, the swelling and shrinkage of the fibers during moisture changes in the composite could affect the fiber–matrix bond. Not only humidity has an influence on the damage process, as the matrix also plays an important role in the process, especially in the vicinity of the fiber. The denser the transition zone is, the higher is the possibility of fiber fracture. When the ITZ is relatively porous, the pull-out behavior is more likely to occur than when the porosity is low. In other words, the presence of dense matrices can prevent the occurrence of the multi-cracking mechanism. Such an occurrence can cause a dilemma for the long-term durability of the material concerned, with the aged matrix tending to compress/squeeze the fiber by means of continuing the precipitation of hydration products around the fiber, making the ITZ denser, and, hence, reducing the probability of microcracking.

The fiber–matrix interfacial transition zone of coconut is presented in Fig. 3.7 after its submission to 25 cycles of wetting and drying (Toledo Filho et al. 2000). In the instance described, the moisture movement has resulted in the mineralization of the fiber leading to the deposition of calcium hydroxide in the lumen and voids of the fiber cells.

Moisture content in fiber influences the degree of crystallinity, the crystalline orientation, the tensile strength, the swelling behavior, and the porosity of fibers.

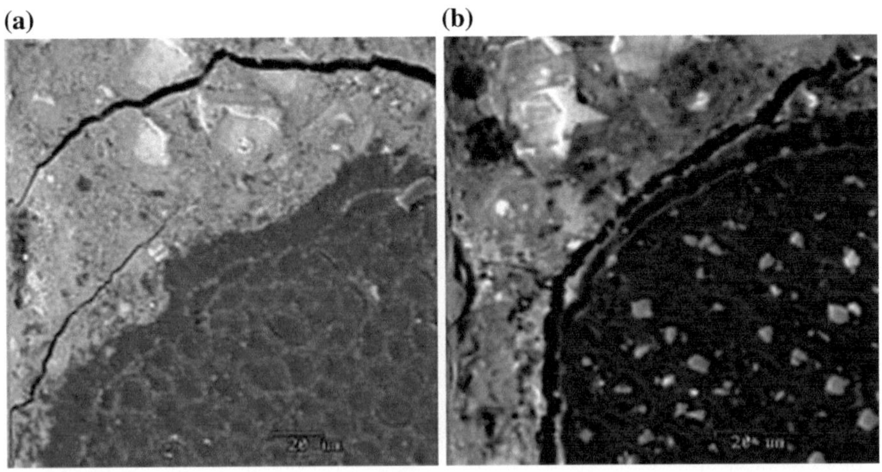

Fig. 3.7 Fiber–matrix ITZ: **a** Coconut fiber–mortar interface zone for a specimen 28 days old; **b** Coconut fiber–mortar interfacial zone for a specimen subjected to 25 cycles of wetting and drying (Toledo Filho et al. 2000)

Relatively high moisture absorption increases the ease of microbial attack (i.e. biodegradation) (John and Thomas 2008).

The moisture absorption of the fiber can be reduced by means of alkaline treatment [i.e. by means of treatment with potassium hydroxide (KOH), or sodium hydroxide (NaOH)]. The treatment is used to decrease the hydrogen-bonding capacity of cellulose, and to eliminate the open hydroxyl groups that tend to bond with water molecules. In addition, the treatment also dissolves hemicelluloses (most hydrophilic), thus reducing the ability of fiber to absorb moisture (Dittenber and Ganga Rao 2012; Symington et al. 2009).

The durability of the natural fibers in the composite material depends on their humidity content. With an increase in humidity, the fibers tend to lose their stiffness, and to become more ductile than before. In addition, their tensile strength decreases. Despite the fibers possibly reaching relatively high ultimate strain value, a reduction occurs in the Young's modulus (Mai and Hakeem 1984; Uzomoka 1976), in comparison with what occurs in the case of low-humidity fibers.

3.6.2 *Thermal Resistance*

Fire degradation is related to two components of natural fiber. A relatively high cellulosic content results in higher flammability than would otherwise be the case, whereas relatively high lignin content results in an increase in char formation. In terms of the fiber microstructure involved, high crystallinity and lower polymerization tend to improve fire resistance. During the thermal decomposition of lignin, relatively weak bonds tend to break at relatively low temperatures, whereas the cleavage of stronger bonds in the aromatic rings takes place at higher temperatures. With lower lignin content, degradation begins at a relatively high temperature, but the fibers concerned lack the oxidation resistance given by the aromatic rings in the lignin (Chapple and Anandjiwala 2010). Figure 3.8 presents the thermogravimetric decomposition process of a natural fiber (i.e. sisal) both under natural conditions, and after being submitted to different cycles of wetting and drying (Melo Filho et al. 2013).

As reported by Melo Filho et al. (2013), most of the natural fiber thermal decomposition (60% of mass fraction) occurs at temperatures ranging between 215 and 310 °C. The thermal decomposition of natural fiber has been divided into three stages (Alvarez and Vázquez 2004; Araújo et al. 2008; Lee and Wang 2006; Manfredi et al. 2006; Suardana et al. 2011), the data relating to which are noted in tabular form, as is shown in Table 3.5.

A microstructural study of the effect of temperature on natural sisal fibers used in cementitious matrices was performed by Melo Filho et al. (2011). For temperatures up to 200 °C, no signs of damage were observed. However, when increasing the temperature to 250 °C, severe degradation to the secondary cell wall of several fiber cells was observed (compare Fig. 3.9a, b). A temperature of 250 °C also caused some damage to the exterior layer of the fiber, as can be seen from the microcrack

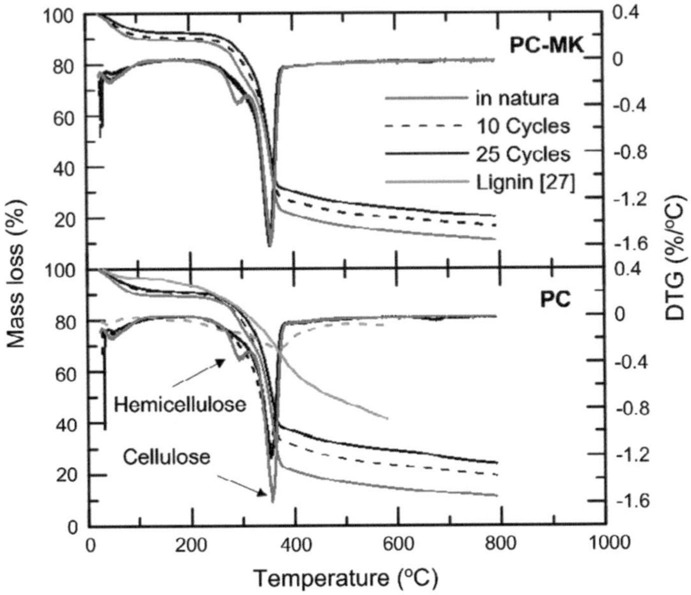

Fig. 3.8 Thermogravimetric decomposition process of natural sisal fiber (Melo Filho et al. 2013)

Table 3.5 Three stages of the thermal decomposition of natural fibers

Stage 1	Stage 2	Stage 3	References
50–100 °C: Evaporation of moisture in the fibers	200–300 °C: Decomposition of hemicelluloses	400–500 °C: Weight loss, due to lignin and cellulose degradation	Araújo et al. (2008)
300 °C: Corresponds to the thermal decomposition of hemicellulose and the glycosidic links of cellulose	360 °C: Corresponds to the thermal decomposition of α-cellulose	200–500 °C, max. at 350 °C: Wider lignin peak appears to be superposed on the other two peaks	Araújo et al. (2008), Manfredi et al. (2006), Alvarez and Vasques (2004)
250–300 °C: Characteristic of such low- molecular-weight components as hemicelluloses	300–400 °C: Corresponds to the thermal degradation of cellulose	Near 420 °C: Due to lignin decomposition	Lee and Wang (2006)
220–315 °C: Pyrolysis of hemicelluloses	315–400 °C: Pyrolysis of cellulose	160–900 °C: Pyrolysis of lignin	Suardana et al. (2011)

formation in Fig. 3.9c. In Fig. 3.9d, it is observed that the constituents of the thick secondary wall (which is the wall that is responsible for the strength of the fiber) are suffering deterioration, probably due to the breakage of the bonds involved. The thick secondary wall consists of cellulose, hemicellulose and lignin. At the micro

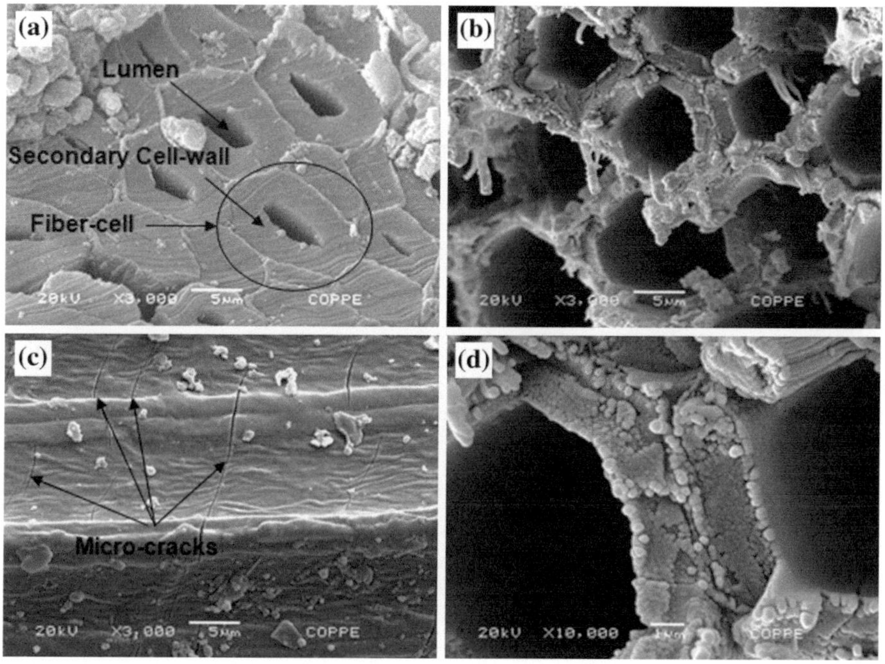

Fig. 3.9 Sisal fiber micrographs after exposure to elevated temperatures inside a cement matrix (Melo Filho et al. 2011)

level, the several crystalline celluloses are linked to one another by means of hemicellulose and lignin. Probably the bond of hemicellulose and lignin, together with the crystalline cellulose, was broken down due to long exposure to temperatures over 200 °C.

3.6.3 Chemical and Biological Degradation

Natural fiber degradation within an alkaline environment was first pointed out by Gram (1983, 1986), who detected two major mechanisms of deterioration: peeling off, and alkaline hydrolysis. The peeling-off mechanism consists of a liberation of the end groups of the molecular chain involved, due to the reaction of the reducing end group and the OH^- ions. The influence of the mechanism is small at temperatures below 75 °C. Alkaline hydrolysis reduces the degree of polymerization concerned (with, for the cellulose molecule, the degree being about 25,000) by means of dividing up the molecular chain. The process has the most influence on the hemicellulose and lignin components involved. Both components are mostly present in the walls of the fibers, so that the degradation breaks the link between individual fiber cells, leading to the loss of its reinforcing efficiency (see Fig. 3.10).

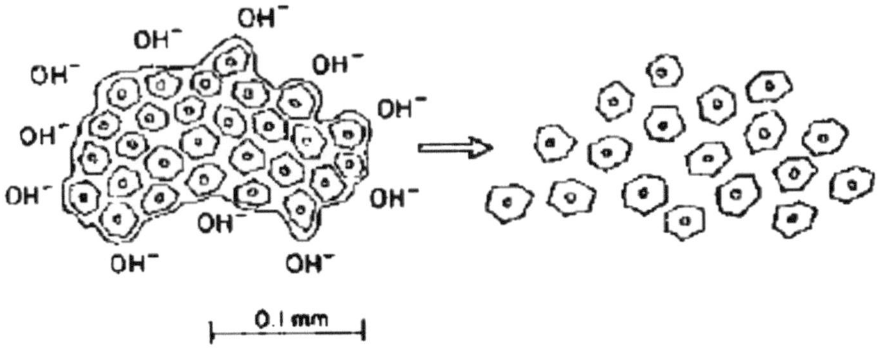

Fig. 3.10 Degradation of sisal fibers in concrete environment, by means of the decomposition and dissolution of the middle lamellae in the alkaline pore water (Gram 1983)

The influence is less significant in pulp fibers from which the middle lamella (consisting of lignin and hemicellulose) is removed during the chemical treatment.

Two natural fiber degradation mechanisms have been proposed by Melo Filho et al. (2013), who used the fibers as reinforcement in Portland cement composites: (i) fiber mineralization due to the precipitation of calcium hydroxide in the fiber cell and surface; and (ii) the degradation of cellulose, hemicellulose and lignin, due to the adsorption of calcium and hydroxyl ions (see Fig. 3.11). The degradation process involved occurs rapidly, with it being possible, after ten cycles of wetting/drying, to observe a quite expressive modification in the composite flexural behavior.

To prevent the fibers from degrading, a silica fume treatment was proposed by Toledo Filho et al. (2003) (consisting of the immersion of fibers in silica fume slurry before casting). The replacement of cement by silica fume also tends to improve the durability by reducing the matrix alkalinity (where fly ash is not effective). Another possibility is to use an alumina cement matrix. Also, treatment of the fibers with water-repellent agents (e.g. formine and stearid acid) is possible, or else with polymer coatings and plasma.

Partial replacement of the cementitious matrix by means of calcined clays, aimed at the reduction of the calcium hydroxide content in a cement-based composite, has been shown, in several works, to be a highly promising treatment (Silva et al. 2010; Toledo Filho et al. 2000, 2009).

Biological attack on fibers can be caused by means of bacteria or fungi that can grow in a humid/moist environment during the wetting/drying cycles, in combination with a warm climate. The alkaline nature of the matrix involved seems to prevent the composite being adversely affected by the problem, in the case of fiber-reinforced concrete (FRC) (Uzomoka 1976). The reduction of the matrix alkalinity could, however, increase the risk of biological attack. In contrast, the small size of the cracks in SHCC could provide sufficient barrier to prevent biological attack.

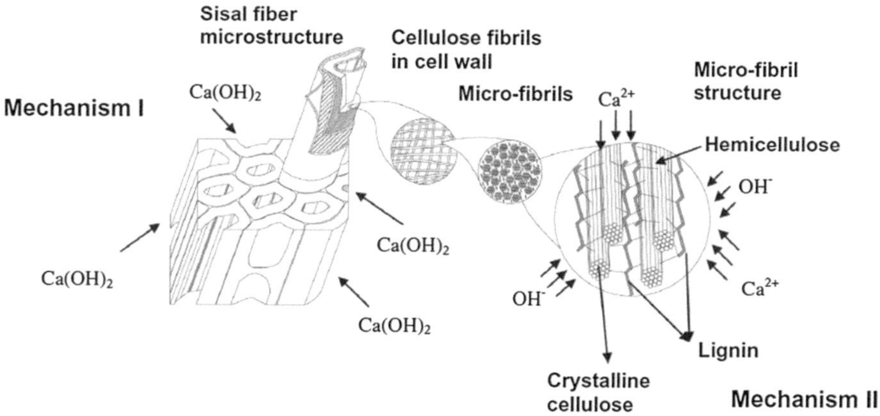

Fig. 3.11 Natural fiber degradation mechanisms inside a Portland cement matrix (Melo Filho et al. 2013)

3.7 Conclusions

The present chapter has reported an overview of the durability of fibers that are commonly used as reinforcement in SHCC materials.

Proper precautions have to be taken into account when using such fibers in aggressive environments. PVA, PP and PE, which are resistant to alkalis, are stable. Nevertheless, the fibers are susceptible to elevated temperatures (of about 180 °C), with their interface possibly being affected at moderate temperatures (60 °C).

Moreover, natural and glass fibers are sensitive to alkaline environments and can undergo severe degradation, if they are used to reinforce Portland cement. Several countermeasures can be used to mitigate this problem, including the partial replacement of cement by pozzolanic materials, fiber surface treatment, and the use of AR glass fibers.

References

Almusallam, A.A. (2001). Effect of degree of corrosion on the properties of reinforcing steel bars. Construction and Building Materials 15(8):361-368.

Alvarez,V.A., Vázquez, A. (2004). Thermal degradation of cellulose derivatives/starch blends and sisal fiber biocomposites. Polymer Degradation and Stability 84:13-21.

Araújo, J.R., Waldman, W.R., De Paoli, M.A. (2008). Thermal properties of high density polyethylene composites with natural fibers: coupling agent effect. Polymer Degradation and Stability 93:1770-1775.

Azwa, Z.N., Yousif, B.F., Manalo, A.C., Karunasena, W. (2013). A review on the degradability of polymeric composites based on natural fibers. Materials and Design 47:424-442.

Balaguru, P.N., Shah, S.P. (1992). Fiber reinforced cement composites. McGraw Hill, New York.

Balouch, S.U., Forth, J.P., Granju, J.L. (2010). Surface corrosion of steel fiber reinforced concrete. Cement and Concrete Research 40(3):410-414.

Banholzer, B. (2004). Bond behavior of a multifilament yarn embedded in a cementious matrix. PhD dissertation, RWTH Aachen, Germany.

Bentur, A. (1989). Silica fume treatments as means for improving durability of glass fiber reinforced cements. Journal of Materials in Civil Engineering 1(3):167-183.

Bentur, A., Diamond, S. (1987). Aging and microstructure of glass fiber cement composites reinforced with different types of glass fibers. Durability of Building Materials 4:201-226.

Bentur, A., Mindess, S. (2007). Fiber reinforced cementitious composites. Taylor and Francis, London.

Bentur, A., Tirosh, R., Yardimci, M., Puterman, M., Peled, A. (2010). Bonding and microstructure in textile reinforced concrete, in Textile reinforced concretes. Proceedings of the International RILEM Conference on Materials Science, Volume 1. W. Brameshuber, editor, RILEM Publications, Aachen, Germany, pp. 23-33.

Bentur, A., Yardımcı, M.Y., Tirosh, R. (2013). Preservation of telescopic bonding upon aging of bundled glass filaments by treatments with nano-particles. Cement and Concrete Research 47:69-77.

Brameshuber, W., Brockmann, T., Banholzer, B. (2006). Material and bonding characteristics for dimensioning and modeling of textile reinforced concrete (TRC) elements. Materials and Structures 39:749-763.

Brandt, A.M., Li, V.C., Marshal, I.H. (1994). Interface strengthening mechanism in polymeric fiber reinforced cementitious composites. Proceedings of the International Symposium Brittle Matrix Composites 4, 13-15 September, Warsaw.

Butler, M., Mechtcherine, V., Hempel, S. (2009). Experimental investigations on the durability of fiber–matrix interfaces in textile-reinforced concrete. Cement and Concrete Composites 31:221-231.

Chapple, S.C., Anandjiwala, R. (2010). Flammability of natural fiber-reinforced composites and strategies for fire retardancy: a review. Journal of Thermoplastic Composite Materials 23 (6):871-893.

Cohen, Z., Peled, A. (2010). Controlled telescopic reinforcement system of fabric-cement composites — durability concerns. Cement and Concrete Research 40:1495-1506.

Cohen, Z., Peled, A. (2012). Effect of nanofillers and production methods to control the interfacial characteristics of glass bundles in textile fabric cement-based composites, Composites: Part A 43:962-972.

De Lhoneux, B., Akers, S., Alderweireldt, L., Amiya, S., Carmeliet, J., Hikasa, J., Saenen, W., Studinka, J., Tomka, I., Van den Bosch, M. (2002). Durability study of PVA fibres in fibre-cement products, Proceedings of the 4th International Symposium of Concrete for a Sustainable Agriculture, Agro-, Aqua- and Community Applications, 21-24 April 2002, Ghent, Belgium, pp. 275-284.

Dittenber, D.B., Ganga Rao, H.V.S. (2012). Critical review of recent publications on use of natural composites in infrastructure. Composites Part A: Applied Science and Manufacturing 43 (8):1419-1429.

Dvorkin, D. (2014). MSc thesis, Ben Gurion University of The Negev, Beersheba, Israel.

Fidelis, M.E.A., Pereira, T.V.C, Gomes, O.F.M., Silva, F.A., Toledo Filho, R.D. (2013). The effect of fiber morphology on the tensile strength of natural fibers. Journal of Materials Research and Technology 2(2):149-157.

Gao, S.L., Mäder, E., Plonka, R. (2004). Coatings for glass fibers in a cementitious matrix. Acta Materialia 52(16):4745-4755.

Glowania, M.H., Linke, M., Gries, T. (2011). Coating of AR-glass fibers with polyurethane for textile-reinforced concrete, 9th International Symposium on High Performance Concrete – Design, Verification & Utilization Energy Events Centre, Rotorua.

Gram, H.E. (1983). Methods for reducing the tendency towards embrittlement in sisal fibre concrete. Publication No. 5, Nordic Concrete Research, pp. 62-71.

Gram, H.E. (1986). Durability studies of natural organic fibers in concrete, mortar or cement. Proceedings of the RILEM Symposium Developments in Fiber Reinforced Cement and Concrete, Sheffield.

Granju, J. L., Balouch, S.U. (2005). Corrosion of steel fiber reinforced concrete from the cracks. Cement and Concrete Research 35(3):572-577.

Gupta, P.K. (1988). Glass fibers for composite materials. In A.R. Bunsell, editor, Fiber reinforcements for composite materials. Composite materials series, Volume 2, Chapter 2, Elsevier, Amsterdam, pp. 20-71.

Hannant, D.J., Zonsveld, J. (1980). Polyolefin fibrous networks in cement based matrices for low cost sheeting. Philosophical Transactions of the Royal Society of London A 294:591-597.

Hoff, G. (1987). Durability of fiber reinforced concrete in a severe marine environment. Concrete Durability-Katharine and Bryant Mather International Conference SP-100, American Concrete Institute, Detroit, pp. 997-1041.

Hoshiro, H., Nishiyama, M., Yamamoto, R. (2006). Long-term durability of Kuralon (PVA fiber) in alkaline condition, Proceedings of the 10th International Inorganic-bonded Fiber Composite Conference.

Hull, D. (1981). An introduction to composite materials. Cambridge Solid State Science Series. Cambridge University Press, Cambridge.

Japan Society of Civil Engineers (JSCE) (2008). Recommendations for design and construction of high performance fiber reinforced cement composites with multiple fine cracks (HPFRCC). Concrete Engineering Series 82.

Jin, W., Zhao, Y., Yan, F. (2001). The mechanism of corroded expansion force of reinforced concrete members. Journal of Hydraulic Engineering 07:57-62.

John, M., Thomas, S. (2008). Biofibers and biocomposites. Carbohydrate Polymers 71 (8):343-364.

Kosa, K., Naaman, A.E. (1990). Corrosion of steel fiber reinforced concrete. ACI Materials Journal 87(1):27-37.

Kuraray. (2015). Standard properties for Kuralon PVA fibers. Retrieved from http://kuralon-frc.kuraray.com/product-application/for-mortar/recs on 09/2015.

Lee, S.H., Wang, S. (2006). Biodegradable polymers/bamboo fiber biocomposite with bio-based coupling agent. Composites Part A: Applied Science and Manufacturing 37:80-91.

Li, V.C., Horikoshi, T., Ogawa, A., Torigoe, S., Saito, T. (2004). Micromechanics-based durability study of polyvinyl alcohol-engineered cementitious composite. ACI Materials Journal 101 (3):242-248.

Magalhães, M.S., Toledo Filho, R.D., Fairbairn, E.M.R. (2010). Physical and mechanical properties of strain-hardening cement-based composites (SHCC) after exposure to elevated temperatures. Advances in Cement based Composites, 17-19 November 2009, CRC Press, Leiden, pp. 203-207.

Magalhães, M.S., Toledo Filho, R.D., Fairbairn, E.M.R. (2015). Thermal stability of PVA fiber strain hardening cement-based composites. Construction and Building Materials 94:437-447.

Mai, Y.W., Hakeem, M.I. (1984). Slow crack growth in cellulose fiber cements. Journal of Materials Science 19:501-508.

Majumdar, A.J., Nurse, R.W. (1974). Glass fiber reinforced cement. Building Research Establishment Current Paper, CP79/74, Building Research Establishment, Watford.

Manfredi, L.B., Rodríguez, E.S., Wladyka-Przybylak, M., Vázquez, A. (2006). Thermal degradation and fire resistance of unsaturated polyester, modified acrylic resins and their composites with natural fibers. Polymer Degradation and Stability 91:255-261.

Masthoff, A. (1998). Analysis using the SIC method and the H-test in alkali resistant glass fibers, E-glass and C-glass fibers, Berlin.

Melo Filho, J.A., Silva, F.A., Toledo Filho, R.D. (2011). Thermo-mechanical behavior of continuous sisal fiber cement based composite systems. Proceedings of the 2nd International Conference on Concrete Spalling due to Fire Exposure, Delft.

Melo Filho, J.A., Silva, F.A, Toledo Filho, R.D. (2013). Degradation kinetics and aging mechanisms on sisal fiber cement composite systems. Cement and Concrete Composites 40:30-39.

Methacanon, P., Weerawatsophon, U., Sumransin, N., Prahsarn, C., Bergado, D.T. (2010). Properties and potential application of the selected natural fibers as limited life geotextiles. Carbohydrate Polymers 82(4):1090-1096.

Nawy, E.G. (1996). Fundamentals of high-strength, high-performance concrete. Addison Wesley Longman, Reading, MA, p. 350.

Oliveira, A., Silva, F.A., Toledo Filho, R.D., Fairbairn, E.M.R. (2014). Temperature and internal moisture effects on the tensile behavior of strain hardening cement-based composites (SHCC) reinforced with PVA fibers. Proceedings of the 3rd International RILEM Conference on Strain Hardening Cementitious Composites (SHCC-3), Dordrecht, RILEM Publications S.A.R.L, Bagneux, France, pp. 51-60.

Oliveira, E.C.P., Lameira, O.A., Sousa, F.I.B., Silva, R.J.F. (2008). Leaf structure of curaua in different intensities of photosynthetically active radiation. Pesquisa Agropecuária Brasileira 43 (2):163-169.

Raupach, M., Orlowsky, J., De Bolster, E., Van Itterbeeck, P., Wastiels, J., Cuypers, H. (2006a). Durability of glass fiber reinforced composites experimental methods and results. Composites Part A: Applied Science and Manufacturing 37:207-215.

Raupach, M., Orlowsky, J., Büttner, T., Dilthey, U., Schleser, M. (2006b). Epoxy-impregnated textiles in concrete – load bearing capacity and durability. Proceedings of the 1st International RILEM Conference on Textile Reinforced Concrete, 6-7 September 2006, Aachen, Germany, pp. 77-88.

Saechtling, H. (1987). International plastic handbook, Hanser Publishers, Munich.

Silva, F.A., Chawla, N., Toledo Filho, R.D. (2008). Tensile behavior of high performance natural (sisal) fibers. Composites Science and Technology 68:3438-3443.

Silva, F.A., Toledo Filho, R.D., Melo Filho, J.A., Fairbairn, E.M.R. (2010). Physical and mechanical properties of durable sisal fiber–cement composites. Construction and Building Materials 24:777-785.

Suardana, N.P.G., Ku, M.S., Lim, J.K. (2011). Effects of diammonium phosphate on the flammability and mechanical properties of bio-composites. Journal of Materials Research 32:1990-1999.

Symington, M.C., Banks, W.M., West, O.D., Pethrick, R.A. (2009). Tensile testing of cellulose based natural fibers for structural composite applications. Journal of Composite Materials 43 (9):1083-1108.

Toledo Filho, R.D., Ghavami, K., England, G.L., Scrivener, K. (2003). Development of vegetable fiber-mortar composites of improved durability. Cement and Concrete Composites 25 (2):185-196.

Toledo Filho, R.D., Scrivener, K., England, G.L., Ghavami, K. (2000). Durability of alkali-sensitive sisal and coconut fibers in cement mortar composites. Cement and Concrete Composites 22(2):127-143.

Toledo Filho, R.D., Silva, F.A., Fairbairn, E.M.R., Melo Filho, J.A. (2009). Durability of compression molded sisal fiber reinforced mortar laminates. Construction and Building Materials 23:2409-2420.

Uzomoka, O.J. (1976). Characteristics of akwara as a reinforcing fiber. Magazine of Concrete Research 28:162-167.

Waweru, R.N. (2011). The effect of fiber corrosion on shear capacity of steel fiber reinforced concrete beams and an initial investigation on alkali–silica reaction in steel fiber reinforced concrete. [M.S.], The University of Texas at Arlington, Texas, United States.

Wong, K.J., Yousif, B.F., Low, K.O. (2010). The effects of alkali treatment on the interfacial adhesion of bamboo fibers. Journal of Materials Design and Applications 224(3):139-148.

Zhu, W., Bartos, P. (1997). Assessment of interfacial microstructure and bond properties in aged GRC using novel microindentation method. Cement and Concrete Research 27(11):1701-1711.

Chapter 4
Chemical Processes

Erik Schlangen, Gideon P.A.G. van Zijl and Petr Kabele

Abstract In this chapter, the influence of various chemical processes on the performance of strain-hardening cement-based materials (SHCC) is discussed. SHCC was found to increase in strength when subjected to long-term elevated temperature and humidity, but to experience a decrease in strain capacity under the same conditions. Also, chlorides tend to decrease the ductility of SHCC when the material is exposed to a chlorine environment for a long time. SHCC, especially when high amounts of fly ash are used in the mix, generally do not experience problems related to alkali–silica reaction (ASR). However, in cases when reactive aggregates are used and the alkalinity in the pore water is high, ASR can occur. The fibres have been found to suppress ASR expansion and, because of the local restraining in the material, the ASR reaction can even be slowed down. The last part of the chapter deals with the self-healing properties of SHCC. Because SHCC is characterised by small cracks it has an intrinsic self-healing capacity. Parts of these cracks close by themselves when the material is stored in a humid environment or under water. Several methods of improving the self-healing capacity of SHCC by adding different types of self-healing agents to the mix are also reported.

Keywords Thermal and moisture cycles · Alkaline environment · Chloride environment · Leaching · Strength and ductility characteristics · ASR · Self-healing

E. Schlangen (✉)
Delft University of Technology, Delft, The Netherlands
e-mail: Erik.Schlangen@tudelft.nl

G.P.A.G. van Zijl
Stellenbosch University, Stellenbosch, South Africa

P. Kabele
Czech Technical University in Prague, Prague, Czech Republic

G.P.A.G. van Zijl and V. Slowik (eds.), *A Framework for Durability Design with Strain-Hardening Cement-Based Composites (SHCC)*, RILEM State-of-the-Art Reports 22,
DOI 10.1007/978-94-024-1013-6_4

4.1 Introduction

The state-of-the-art of strain-hardening cement-based materials (SHCC) durability under chemical attack was reported by Oh and Kabele (2011), who discuss the effects of chloride on steel reinforcement embedded in SHCC, the matrix, and the matrix–fibre interface. They also describe the effects of hydrolysis and leaching, hot and humid conditions, sulphate attack, and alkali–silica reaction (ASR) on SHCC. Significant research on several of the topics has since been performed and reported. The corrosion of steel reinforcing bars in SHCC has received renewed attention due to the concern raised in the previous State-of-the-Art Report (STAR) (Van Zijl and Wittmann 2011) that fine cracks in SHCC act as pathways for the quick ingress of water and chlorides by capillary absorption. Chapter 9 of the current report is devoted to corrosion in steel-reinforced SHCC (R/SHCC), due to the new level of understanding attained and importance of this deterioration process. ASR in fibre-reinforced concrete and SHCC has also been the subject of continued research. New insights have been gained into the mechanisms of delay and lower reactivity of ASR by bridging ASR-induced cracks with fibre. Significant new research results on the autogenic and autonomic healing of fine cracks in SHCC have also been produced and reported on since the previous STAR. In this chapter, the effects of aging under different conditions, ASR, and self-healing in SHCC are presented.

4.2 Aging Under Various Moisture, Thermal and Chemical Conditions

Several experimental studies have been carried out to investigate the performance of SHCC materials and their constituents under conditions representing accelerated aging in different environments.

Li et al. (2004) investigated the effect of a hot and humid environment on polyvinyl alcohol (PVA)–SHCC and its constituents. The exposure was realised by means of immersing mature specimens in 60 °C water for 0, 4, 13, 26 and 52 weeks. Individual PVA fibres showed almost no change in nominal strength, elastic modulus, and elongation. Although fibre pull-out tests revealed that the frictional bond at the fibre–matrix interface was almost unaffected by the exposure, notable increase of the chemical bond and decrease of the apparent fibre strength were observed after 13 weeks of accelerated aging. Uniaxial tests on composite specimens showed that the first crack strength and the ultimate tensile strength increased with time of exposure. Despite all specimens exhibiting multiple cracking, the tensile strain capacity had a decreasing tendency, especially from 13 to 26 weeks after exposure. Qian and Zhang (2012) explore the possibility of utilising accelerated curing at high temperatures and humidity for rapid quality control of PVA–SHCC. They cured fresh composite specimens in an environment with temperature of 75 °C and with relative humidity of 98% for two days, whereas control samples were kept at 20 °C and 98% RH for 28 days. Both series were then tested in four-point bending, from which tensile properties were extracted by means of inverse analysis. The specimens that were exposed to accelerated curing exhibited slightly higher tensile strengths, but lower strain capacity than did the others. Ibrović et al. (2014) examined the aging of SHCC with PVA fibres and matrix, which contained, besides the usual constituents, limestone powder and blast-furnace slag. The tests consisted in placing thin (3 and 5 mm) specimens in an oven that was preheated to 70 °C for five days, which was followed by storing them at 25 °C and at a relative humidity of 95% for two days, after which the specimens showed no cracking. Subsequently, the specimens were loaded into four-point bending. Compared to the specimens that were cured at room temperature, the exposed samples showed lower ductility and higher first cracking strength. Sierra Beltran et al. (2014b) studied the durability of PVA fibre-reinforced SHCC that contained either coal or biomass FA. The presence of a harsh environment was simulated, among other simulations, by means of submersion into a 80 °C water bath (I), by means of curing in a dry environment at 20 °C and 50% RH (III), and by means of exposing the specimens to wet–dry cycles at 20 °C (IV). Compared to the moist-cured control specimens, the average compressive strength decreased after exposure (I), and even more notably after (III), but it was almost unaffected by wet–dry cycles (IV). Bending tests on thin plates revealed that the hot water bath (I) caused an increase of the first crack strength, but reduced the ultimate MOR and deflection capacity. The wet–dry cycles (IV) had almost no effect on the bending response.

Şahmaran and Li (2008) experimentally investigated the influence of a high alkaline environment on PVA–SHCC with non-reactive silica sand. The researchers in question subjected virgin (unloaded) SHCC specimens, as well as specimens that had been mechanically preloaded to cause cracks about 50 μm wide, to sodium hydroxide aqueous solution at 38 °C for a period of time up to 90 days long. Subsequently, the samples were tested under conditions of uniaxial tension. Recovery of initial elastic stiffness and the self-healing of cracks were reported for the pre-cracked specimens. Both uncracked and cracked specimens showed sustained strain hardening behaviour, with average crack widths being in the range of 74–98 μm, compared with the average crack widths in the range of 39–58 μm in similar SHCC specimens of the same age, but that were air-cured after seven days of moist curing. The alkaline environment also caused a roughly 20% reduction of ultimate tensile strain, and a 4% reduction of tensile strength. The phenomena concerned are believed to have been due to a change in the fibre–matrix interface in the alkaline solution. The authors also performed standard (ASTM C-1260) accelerated ASR tests, resulting in no expansion after a 30 day exposure period, probably due to the use of non-reactive silica sand.

Şahmaran and Li (2009) studied the durability of PVA–SHCC with a high content of class F FA. They applied accelerated aging to intact and pre-cracked composite specimens by means of imposing continuous exposure to 1 N sodium hydroxide solution at 38 °C up to 90 days, and 3 N sodium chloride solution at room temperature up to 180 days. Direct tension tests revealed that the alkaline environment caused negligible decrease in the ultimate tensile strength, a decrease of the tensile strain capacity, and an increase in crack width. Specimens stored in the chloride solution showed a slight reduction in their ultimate tensile strength over the 180-day exposure period. The tensile strain capacity did not appear to be affected within the 90-day exposure period, but it decreased by up to 30% after the course of 180 days. The healing of cracks was reported in both the environments concerned. Kabele et al. (2007) and Kabele and Pekař (2009) experimentally investigated the effects of chloride exposure, combined with thermal cycles (S-series), and the effects of calcium leaching (N-series) on the mechanical behaviour of PVA–SHCC on different scales. The former represented aging experienced in, among others, marine and road structures, whereas the latter corresponded to conditions of underground or agricultural structures, dams or sewers, which can be attacked by soft water or by nitrate-containing water. All specimens were first allowed to harden under room conditions for 28 days. Subsequent treatment of the S-series consisted of ten cycles of five-day immersion in a saturated solution of NaCl at 20 °C and two days of oven drying at 50 °C. Leaching of the N-series was performed by means of immersing the samples for 70 days in 6 mol/l water solution of NH_4NO_3 at room temperature. Mechanical testing involved uniaxial compression and uniaxial tension tests, 3-point bending on notched beams, and single-fibre pull-out. The study revealed that the combined chloride and thermal exposure resulted in an increase of the first cracking strength and of the fibre–matrix bond, whereas leaching caused their reduction. The tensile strain capacity, however, was significantly lower for the S-series, whereas it was significantly higher for the N-series when the capacity was compared to that of the control specimens kept at room conditions.

From the reviewed research, it can be concluded that the aging of PVA–SHCC at elevated temperature and humidity causes an increase of the strength characteristics across different scales (matrix, fibre–matrix interface, and composite), although it also results in the overall embrittlement of the composite. The presence of chlorides in the process seems to have little effect in the initial stage (up to tens of days), but it does tend to lead to the deterioration of ductility after relatively long exposure. The leaching environment, in contrast, tends to reduce the strength characteristics and to improve the overall ductility of the composite concerned. The findings made are consistent with the multiscale fracture-mechanics-based interpretation of multiple-cracking phenomena (Kabele 2007, 2009): higher fibre–matrix bond strength and matrix fracture resistance tend to result in increased load acting on the fibre bridging involved, which causes the premature rupture of fibres. Such rupture, in turn, causes an increase in the brittle response of individual cracks, which tends to lead to reduced overall ductility, or, even, to loss of the multiple-cracking ability of the composite.

4.3 Alkali–Silica Reaction

ASR is an expansive chemical deterioration process that is caused by the chemical reaction between reactive aggregate and the inherently alkaline cement-based composites. Symptoms of ASR include volume increase, the mapping of crack pattern formation, gel seeping from cracks, and pop-outs of aggregates. Apart from the unsightly appearance that results from the presence of ASR, it can also cause significant deterioration in the mechanical properties of especially stiffness (i.e. Young's modulus) and strength. Crack formation typically presents access to deleterious substances, which subsequently ingress into the matrix, and which may lead to various other deterioration processes, such as to the chloride-induced corrosion of embedded steel reinforcement.

4.3.1 ASR in SHCC

Rokugo et al. (2013) studied ASR swelling in SHCC, by means of the use of reactive sand. The researchers purposefully developed ASR-swelling SHCC to counter drying shrinkage-induced volume reduction for specific applications where shrinkage may be detrimental. Accelerated ASR expansion of up to 0.266% was recorded in the specimens concerned, compared with volume reduction (i.e. shrinkage) in control specimens containing non-reactive silica sand. The specimens were exposed to an accelerated ASR environment by means of immersion in a 1 mol/L NaOH aqueous solution at 60 °C for up to 44 days, after which they were left to dry for 14 days. The SHCC mixes contained JIS R5210 high-early-strength Portland cement and limestone powder, and no FA. The results of flexural tests performed on the specimens, after the ASR exposure and drying periods, showed a 19% reduced ultimate flexural load and significantly lower deflection-hardening trajectory for the ASR-affected SHCC, compared to the control SHCC.

4.3.2 Crack-Control Mitigation of ASR

Bektas et al. (2006) postulated that crack-bridging by fibres mitigates the extent of ASR damage caused. They performed accelerated ASR experiments, according to ASTM C-1260 (2007), on steel microfibre-reinforced concrete specimens. Prism specimens of dimension 40 × 40 × 160 mm containing no fibres (0%), and fibres at volume percentages (V_f) of 1, 3, 5 and 7%, respectively, of brass-coated, straight steel fibres of 6 mm length and of 0.16 mm diameter, were stored in sodium hydroxide solution at 80 °C for 30 days. During the 30 days concerned, the specimen length was periodically monitored. The same matrix and reactive aggregate was used in each case. The specimens containing fibres showed significantly reduced expansion

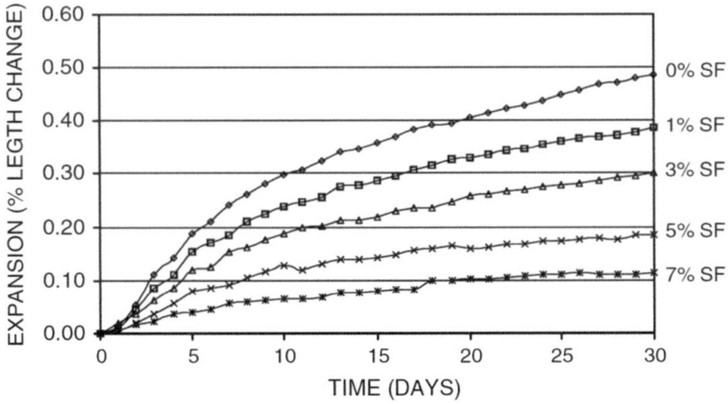

Fig. 4.1 Expansion development during accelerated ASR exposure (Bektas et al. 2006)

compared with the specimens containing no fibre. After 30 days of alkaline exposure, the non-fibrous specimens had expanded by roughly 0.48%, whereas the expansion was 0.12% in the case of $V_f = 7\%$, with intermediate values (0.38, 0.30 and 0.18% respectively for $V_f = 1$, 3 and 5%) being attained for the other fibre volumes (see Fig. 4.1). Note that no characterisation of tensile mechanical behaviour was performed, so it is unknown whether or not the presence of the fibres led to the strain-hardening behaviour involved. However, the mechanism of splitting crack-control is of interest, with it being believed to be relevant to SHCC, should an expansive process occur in its matrix. Scanning electron microscope (SEM) images not only revealed cracks in the aggregates, but also in the matrix of the non-fibrous specimens. In the fibre-reinforced specimens, cracks could not be detected in the matrix, which is ascribed to the microfibres controlling the crack widths and to them preventing the ASR gel from leaving the aggregates concerned.

Similar accelerated ASR tests on steel microfibre-reinforced concrete with $V_f = 7\%$, and on non-fibrous control specimens, were reported by Yi and Ostertag (2005) and by Ostertag et al. (2007). Instead of reactive aggregate, a 5-mm-diameter reactive Pyrex rod was embedded centrally in each specimen. The specimens had a cross-section of 25 × 25 mm, and a length of 280 mm. After curing, the specimens were submerged in NaOH solution at 80 °C, according to ASTM C-1260 (2007). SEM images taken after 42 days of exposure are shown in Fig. 4.2. The dark regions are the reaction rims, in which the ASR reaction products formed. By comparison with Fig. 4.2a, b, the ASR reacted area is significantly reduced by the presence of fibres. In Fig. 4.3, the reacted area is plotted as a function of the reaction time for both the fibrous and the non-fibrous specimens. A clear delay in ASR product formation could be observed for the fibre-reinforced specimens, as well as reduced reactivity, when they were compared with the control specimens. The authors argue, on the basis of the schematic illustration in Fig. 4.3b, that the fibrous crack-control prevented ASR gel from escaping the reaction site, resulting in an increase in the reaction Si ion concentration. The result was reduced dissolution of the reactive aggregate, leading to a reduced ASR rate.

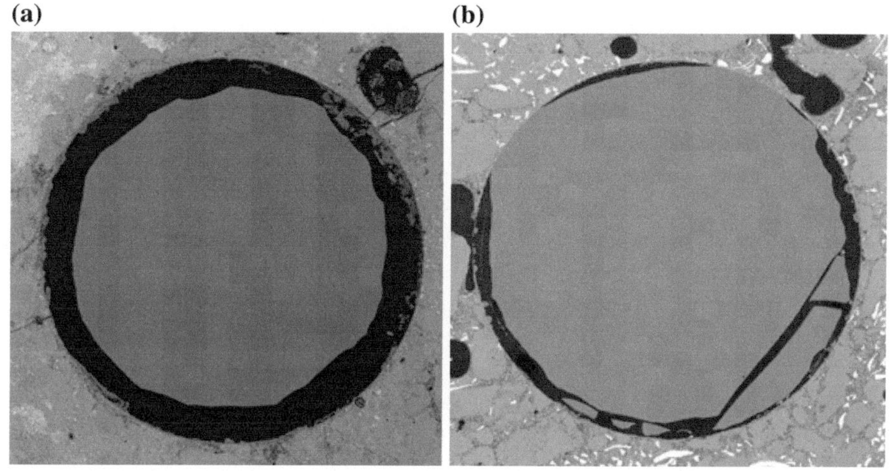

Fig. 4.2 Backscattered (SEM) images of the remaining cross-section of the original 5-mm-diameter Pyrex rod embedded centrically in the specimen, after 42 days of submergence in NaOH at 80 °C with prism specimens containing **a** no fibres, and **b** steel microfibres that are visible as white regions in the matrix (Yi and Ostertag 2005)

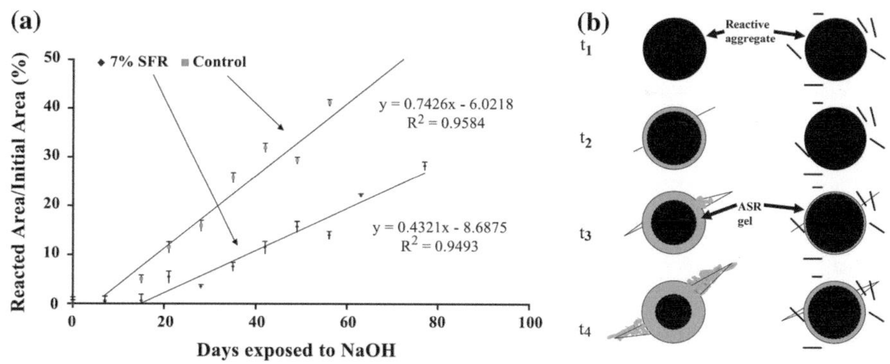

Fig. 4.3 **a** ASR reacted area as % of original rod area for fibrous and non-fibrous specimens (Yi and Ostertag 2005); **b** Schematic illustration of reduced reactivity by means of fibre crack-control (Ostertag et al. 2007)

4.3.3 Fibre Pull-Out from ASR-Affected Cement-Based Matrix

Single-fibre pull-out from ASR-affected matrices was performed by Beglarigale and Yazici (2013). Hooked-end steel fibres of a length of 60 mm and of a diameter of 0.75 mm were embedded in seven different matrix compositions, with main parameter cement (CEM I 42.5 R) replacement levels with silica fume (SF), slagment (GGBS) and FA. Reactive basalt aggregate with particle sizes 1–4 mm was used in the mixes to prepare 50-mm-side-length cubes. A single fibre was embedded centrally in each cube with a 30 mm (half the length of the fibre) embedded length. Accelerated test conditions, according to ASTM C-1260 (2007) (consisting of NaOH kept at 80 °C), were applied to the specimens, after which pull-out of the fibres was conducted. Control specimens of the exact same mixes were prepared, but exposed to water at 80 °C. The difference in the effect of the ASR exposure versus the hot water exposure on the fibre pull-out resistance was marked in the case of matrices with no cement replacement, but they were less marked in cases of significant cement replacement. In the specimens containing only CEM I 42.5 R as binder, the pull-out resistance after one day of ASR exposure was more or less unaffected (−2% less), in comparison with the resistance of the control specimens. After 28 days, a 28% reduction in pull-out resistance was found, but specimens that were subjected to 90 and 150 days of ASR exposure had significantly higher (+59%) fibre pull-out resistance and toughness (+25%) than did their counterparts exposed to hot water for the same time periods. For matrices containing significant amounts of cement replacement materials (FA, SF, GGBS), little or no difference was observed. The authors ascribed the higher pull-out resistance to ASR gel congestion, as was also observed in SEM images of the specimens involved. Significant accumulation of ASR gel in the pores led to congestion and fibre-confining pressure, leading to relatively high pull-out resistance. In matrices containing cement replacement materials, ASR activity is well-known to be reduced. Such reduction explains why no significant change in pull-out resistance was brought about by exposure to ASR-inducing conditions in the specimens containing significant amounts of cement replacement materials in the experiments in which they were used.

4.3.4 ASR in SHCC—Summary

Limited experimental research has been executed on ASR deterioration in SHCC. However, ASR swelling was demonstrated in SHCC containing highly reactive aggregate and with a high total alkali content. In experimental programmes that led to ASR swelling in fibre-reinforced and in control non-reinforced matrices, two important mechanisms have been illustrated. The first is the delayed onset of the ASR chemical process and reduced reactivity, due to the fibre-bridging control of ASR-induced splitting cracks, resulting in the ASR product being prevented from escaping from the reaction site. The second is the increased fibre pull-out resistance from an ASR-affected matrix, due to the congestion and associated pressure build-up in the matrix, which causes higher frictional resistance in the fibre–matrix interface. The mechanisms concerned have been demonstrated in steel-fibre-reinforced cement-based composites including reactive aggregate or reactive Pyrex rods. It remains to be shown whether such ASR-mitigating mechanisms are, indeed, activated in SHCC.

4.4 Self-Healing of SHCC

4.4.1 What is Self-Healing?

Self-healing materials, as materials that can repair their own damage, are a research area that has garnered much attention over the last decade. Several international conferences specific to the topic have already been organised, and it is also a theme that is often found on the topic list of conferences on concrete and other cement-based materials. Furthermore, various textbooks and review papers on self-healing materials have been published in recent years. The RILEM (the Technical Committee of the International Union of Laboratories and Experts in Construction Materials, Systems, and Structures) published a STAR on the self-healing of cement-based materials (De Rooij et al. 2013). In the current STAR, definitions for self-healing are given as follows:

Self-healing Any process undertaken by the material itself, involving the recovery and, hence, the improvement of performance after an earlier action that had reduced the performance of the material.

Autogenic The self-healing process is autogenic when the recovery process uses material components that could otherwise also be present even when the material is not specifically designed for self-healing (i.e. using its own generic materials).

Autonomic The self-healing process is autonomic when the recovery process uses material components that would otherwise not be found in the material (i.e. engineered additions).

In the current chapter, the focus is on the self-healing properties of SHCC. Sometimes the concepts involved have first been tried on general cement-based materials and concrete, as discussed in De Rooij et al. (2013), with them, only after that, having also been applied to SHCC. Both autogenic and autonomous self-healing concepts for SHCC are discussed.

The reason for making concrete materials and SHCC self-healing are to:

- stop the leakage of liquids or gases through cracks;
- block the ingress of liquids or gases into the material, for instance to protect reinforcement bars from corroding, and
- regain such mechanical properties as stiffness, strength, and/or strain capacity.

In the discussion of the various self-healing methods applied to SHCC in the following paragraphs, the main reason for self-healing that is tackled by the specific method is mentioned. In the STAR on self-healing (De Rooij et al. 2013), different techniques are discussed for evaluating the self-healing performance of cement-based materials. The techniques that are applied are mechanical tests, permeability tests, optical microscopy, scanning electron microscopy with energy-dispersive X-ray (EDX), X-ray diffraction (XRD), Fourier transform infrared (FTIR), X-ray computer tomography (CT-scanning), and others. The techniques mentioned are also applied when studying the self-healing of SHCC. Although they are not further discussed in the present text, more information can be found in the specific papers given in the references, or in De Rooij et al. (2013).

4.4.2 Autogenic Self-Healing

To obtain autogenic self-healing in cement-based materials, the requirements are that: (1) the material contains only small cracks; (2) there is material that can be precipitated into the crack; and (3) water is present. SHCC is developed to accommodate small cracks, with the material mostly containing unhydrated cement, or other hydration products, left after the first hydration that can later, when the material is cracked, be used for self-healing. Furthermore, the precipitation of $CaCO_3$ can, and does, take place in accordance with the following reaction:

$$Ca(OH)_2 + CO_2 \rightarrow CaCO_3 + H_2O. \quad (4.1)$$

Water is often also present, especially in cases where the main reason for obtaining self-healing is stopping its leakage or ingress.

Many studies on the self-healing of SHCC deal with autogenic self-healing. As SHCC is developed to be strain-hardening, it is studied with respect to its self-healing properties. In its testing, specimens are mostly loaded up to a certain strain, after which they are exposed to different curing regimes, so as to obtain healing. The curing can occur under water, in air with a certain relative humidity, or with wet–dry cycles. After curing, the specimens are either studied to observe crack filling, or they are retested mechanically, or for permeability, moisture ingress, or diffusivity. Examples of such studies can be found in Fan and Li (2015), Homma et al. (2009), Kan and Shi (2012), Li and Yang (2007), Qian et al. (2010), Schlangen and Joseph (2009), and Ying-zi et al. (2005).

Several investigations have been performed on the healing capacity of SHCC in a marine environment, or in the presence of chloride ions from de-icing salts (Li and Li 2011; Li et al. 2007). Ozbay et al. (2013) performed a study of SHCC under sustained flexural loading, in terms of its effect on self-healing performance. Qian et al. (2009a, b) focus on comparing the self-healing properties of regular SHCC and of SHCC made with different waste materials.

In all the above-mentioned studies, the main self-healing mechanism is precipitation of $CaCO_3$ and/or the ongoing hydration of the unhydrated part of the cement particles concerned. The exact mechanism involved depends on the boundary conditions in the tests. For instance, submersion of the specimens in water leads to different healing products than in the case of the specimens being subjected to wet–dry cycles. In most articles that have so far been published the real healing mechanism involved is not reported on in detail. Furthermore, several researchers assert that the fibres contained in the SHCC, which bridge the crack, work as nucleation sites for the hydration products that are formed by the ongoing hydration (see, for instance, Nishiwaki et al. 2012).

Figure 4.4 shows specimens with dimensions of 120 × 30 × 10 mm that were loaded in four-point bending, up to a deflection of 2.4 mm (Qian et al. 2009b). Thereafter, they were stored under water to heal. After reloading (Fig. 4.4b), some of the healed cracks were observed to reopen, but new cracks were also formed. In general, the stiffness of the specimens can be stated as being restored after healing, with the flexural strength also reaching similar values upon reloading. However, this is a normal feature of SHCC. As new cracks are also formed between existing cracks during the process of reloading, the healing products concerned are also likely to have a strength that is close to that of the original material. However, cracks have been mentioned as staying partly open during healing, as well as has the fact that an existing (healed) crack is most likely to be less strained upon reloading, thus meaning that the stress that requires transferral by the healing product also tends to be lower than before.

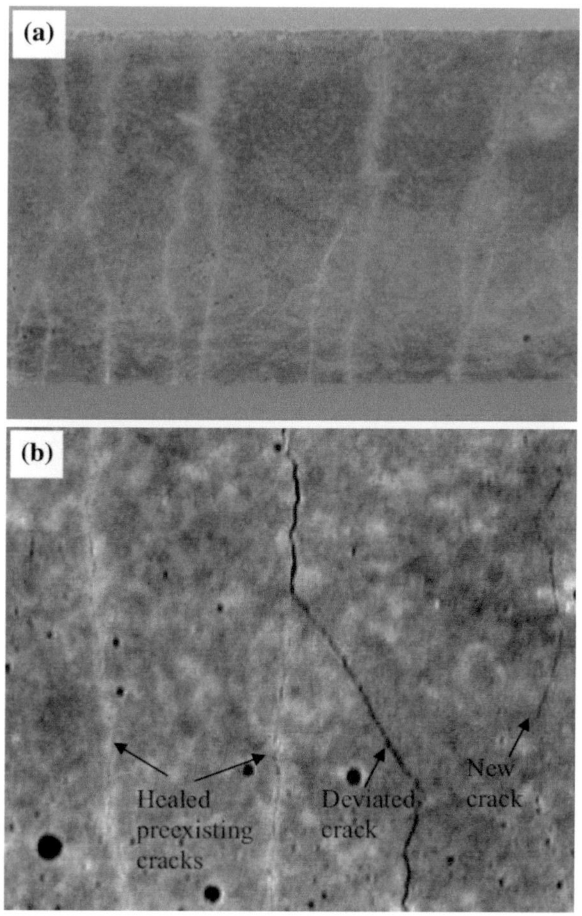

Fig. 4.4 Self-healed specimen **a** before and **b** after retest (Qian et al. 2009b)

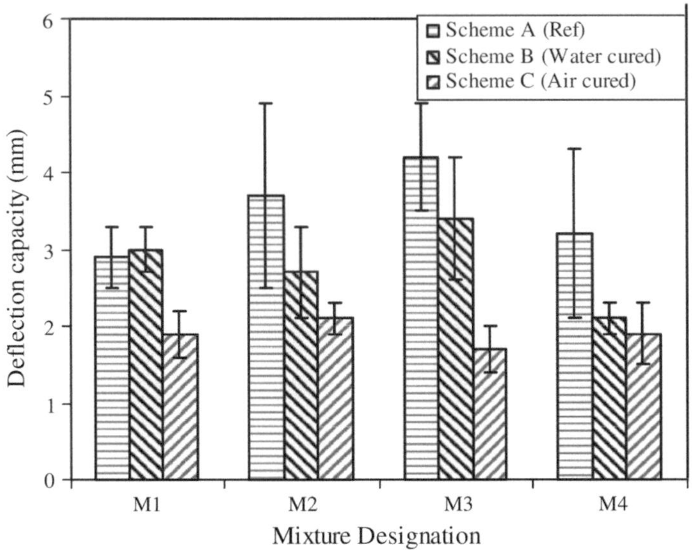

Fig. 4.5 Comparison of the deflection capacity of different mixtures with different curing regimes (water and air) for healing (Qian et al. 2009b)

The high strain capacity of SHCC is the main characteristic of the material. In Fig. 4.5, the strain capacity of reference specimens (Scheme A) is compared with that of specimens that are first cracked, and then healed and reloaded. The healing took place under water (Scheme B) and in air (Scheme C). The tests were performed for four different mixes (Qian et al. 2009b). The recovery of deflection capacity was shown to take place, especially for mix M1, which healed under water. Such recovery of deflection capacity suggests that the water curing involved had considerably promoted the self-healing process, and, therefore, it had enhanced the fibre-bridging behaviour after pre-cracking, whereas the same was not the case for the air-cured pre-cracked samples.

Figure 4.6 shows an ESEM observation of a cracked SHCC sample that was healed under water (Qian et al. 2009b). The small cracks in the range of 15 μm are completely filled with hydration products. The black line is a PVA fibre bridging the crack. Figure 4.7 shows a sample with relatively wide cracks of up to 60 μm. The figure shows that the small cracks (of a width of about 25 μm) are filled with hydration products, but that the relatively wide cracks are only partially filled, which also affects the transport properties involved. If the cracks are only partially filled, they might not block transport. Depending on the curing regime, and whether they are wet or dry, or wet–dry cycles, the applied strain, the age at healing, and the mix composition, different results are reported for the crack-sealing capabilities of SHCC. The results, which are not always reliable, have inspired researchers to look for additions to the SHCC mix to promote self-healing, and create more favourable conditions that are conducive to self-healing.

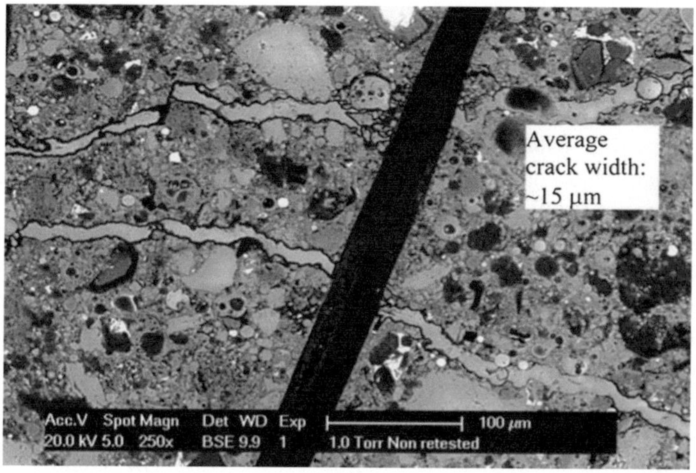

Fig. 4.6 Pre-cracked sample after water curing (no reloading applied) (Qian et al. 2009b)

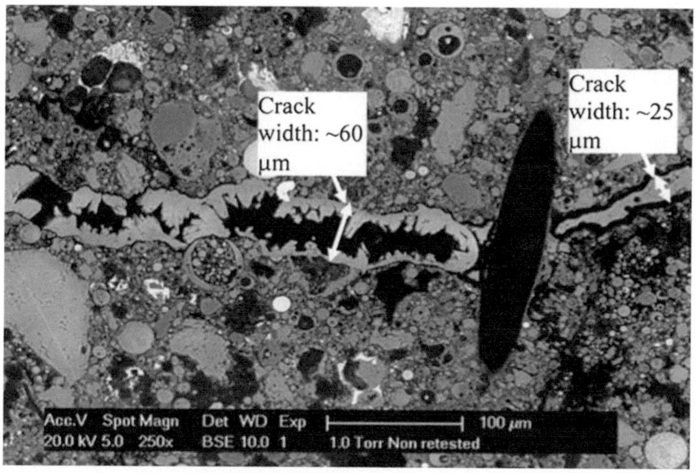

Fig. 4.7 Partial crack healing after water curing (Qian et al. 2009b)

4.4.3 Self-Healing by Means of Adding Extra Water

In Sect. 4.4.2, the self-healing of SHCC was mentioned as not functioning well if the curing takes place in a dry environment. Furthermore, it was observed that, sometimes, proper crack filling takes place only at the surface of a specimen, rather than inside the crack, due to the partial blocking of the ingress of water. For these reasons, inserting additional water into the material, with the water concerned then being able to be used for ongoing hydration after cracks are formed, has been the subject of study by several different research groups.

De Rooij et al. (2009) and Qian et al. (2009a) proposed and tested a method of adding extra water to SHCC by means of using hollow wood fibres, as is shown in Fig. 4.8. The fibres concerned could also be used to add a healing agent (e.g. a superglue) that could heal cracks. A similar study was performed by Ferrara et al. (2014). Schlangen et al. (2009) developed a simulation tool for optimising the amount of healing fibres that need to be added to SHCC to enable the provision of enough water or healing agent to fill cracks of a certain width.

An alternative method for adding extra water to SHCC to promote self-healing entails the addition of superabsorbent polymers (SAPs), or hydrogels. The feasibility of using such a method was first tested by Antonopoulou (2009) and Tziviloglou (2009). Later, Kim and Schlangen (2010, 2011) studied the technique further, finding that it especially showed good performance when wet–dry cycles were applied to the cracked and healing SHCC material. The situation tested was close to that of practical situations. During wetting, the cracked SHCC sucks up water both in the cracks and inside the SAP. During drying cycles, the water from the SAP is slowly released, thus making it available for hydration of the cement, which means that healing can occur even in dry periods. Xia Hua (2010) used encapsulated SAP to avoid the creation of relatively large voids that occurs when SAPs are merely mixed into the fresh SHCC. In the given case, the polymer might

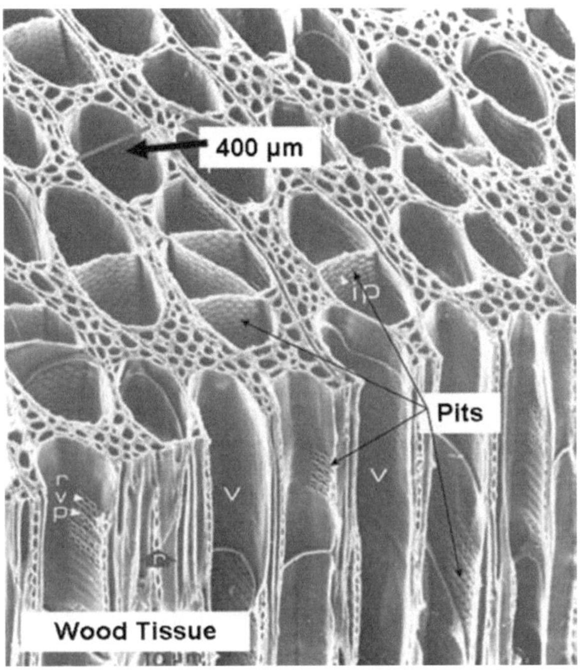

Fig. 4.8 Lignified fibres as carriers for extra water, or for self-healing agent (De Rooij et al. 2009)

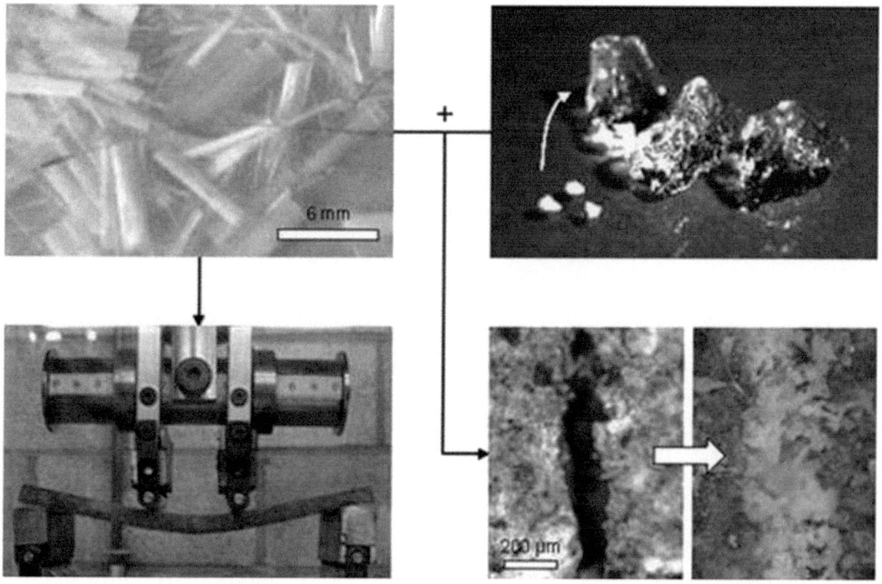

Fig. 4.9 Self-healing of SHCC with PVA fibres and superabsorbent polymers (Snoeck 2015)

also be able to fill a relatively large crack area when the polymer swells after crack formation and the sucking up of water.

In Snoeck's (Snoeck 2015; Snoeck et al. 2012, 2015) highly structured analysis of the use of SAP or hydrogels to promote self-healing in SHCC, different mixes and curing regimes were investigated by means of multiple experimental techniques to obtain evidence of the presence of the self-healing mechanism. The principle of using SAP in SHCC is schematically presented in Fig. 4.9.

4.4.4 Adding (Extra) SH Agent

Another approach to autonomic self-healing entails adding a healing agent to the SHCC mix that promotes self-healing. The addition could both be made for the purposes of crack filling or sealing, or for the regaining of mechanical properties. In the current section, four different approaches that are applied for SHCC are discussed.

4.4.4.1 Fibres

In the first approach to adding SH agent, the agent is applied by means of the use of hollow fibres. Li et al. (1998) tested the principle underlying the method by using

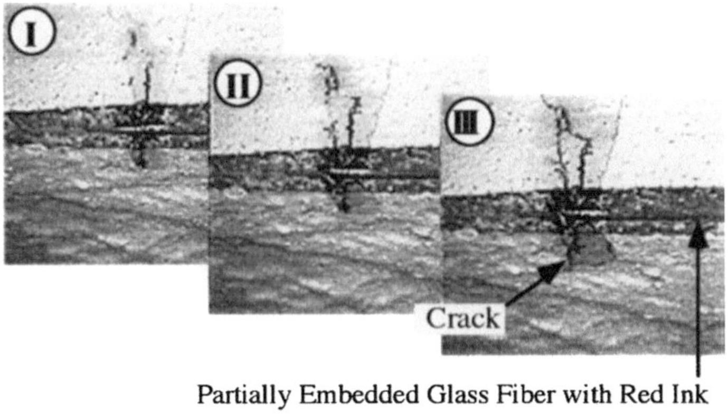

Fig. 4.10 The self-healing of ECC with embedded glass fibre (Li et al. 1998)

red-ink-filled glass tubes (Fig. 4.10). The tests were undertaken to see whether, when a strain is applied to SHCC, cracks will occur, and whether the embedded fibres involved would be stretched and pulled out of the matrix giving the ductility. By means of adding such brittle fibres as hollow glass tubes, the fibres concerned were found to be capable of breaking and releasing the healing agent.

Similarly, De Rooij et al. (2009) used natural hollow wood fibres that were well bonded with the cement matrix, and which had a relatively low strength. The properties involved meant that the fibres would break when the SHCC cracked, and that they would thereby release the healing agent involved.

Antonopoulou (2009) and Tziviloglou (2009) combined PVA fibres, microfibres (Basalt) and SAP, which tended to decrease the crack width in the SHCC, and also to work as nucleation sites. The SAP added extra water for the hydration of the cement, as was previously described.

4.4.4.2 Bacteria

Bacteria have been used in concrete to promote self-healing. An overview of different methods of such bacterial self-healing that have also led to various applications, is given by Tziviloglou et al. (2016). One of the applications concerned uses bacteria in an SHCC repair material (Sierra Beltran et al. 2014a, b, 2015). The bacteria are added to the SHCC as spores that are embedded together with food inside porous sand particles. Once the SHCC cracks, and the pH drops below ten, the bacteria become active, converting the food. In this way, calcium carbonate is precipitated into the crack. By means of the mechanism, the crack is blocked, and the ingress of water and aggressive ions stopped. In Fig. 4.11, an environmental scanning electron microscope (ESEM) image, with an example of the calcium carbonate minerals precipitated in a crack within SHCC, is shown. Li and Herbert (2012) also

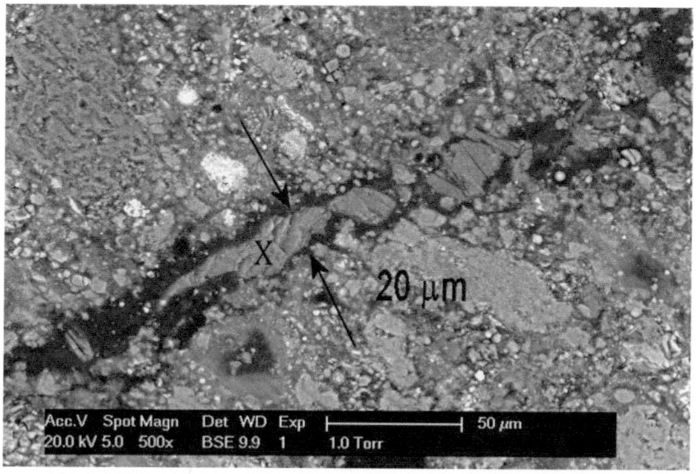

Fig. 4.11 Photomicrograph of internal healing products in bacteria-based SHCC sample (Sierra Beltran et al. 2014c)

added bacteria and nutrients to SHCC, with the aim of studying their effect on its performance.

4.4.4.3 Expansive Agents

Adding expansive agent to concrete is a method of healing cracks that was applied by Sisomphon et al. (2011, 2013) to SHCC. An expansive additive, based on calcium-sulfoaluminate, was used in the research. Various exposure conditions and mix designs were tested. Both crack blocking (see Fig. 4.12) and the recovery of mechanical properties were observed.

4.4.4.4 Shape Memory Alloys

Shape memory alloys (SMAs) are used in materials, as they are able to close cracks after they have formed, because the stretched SMA tends to revert to its original length. When SMAs are used in reinforced concrete, large cracks still tend to be present after unloading. If the technique is applied to SHCC, as was done by Li et al. (2015), the already small cracks usually close completely, which enhances the self-healing capability of SHCC. As shown in Sect. 4.4.2, small cracks in SHCC may be completely filled and healed. Figure 4.13 illustrates the SMA principle.

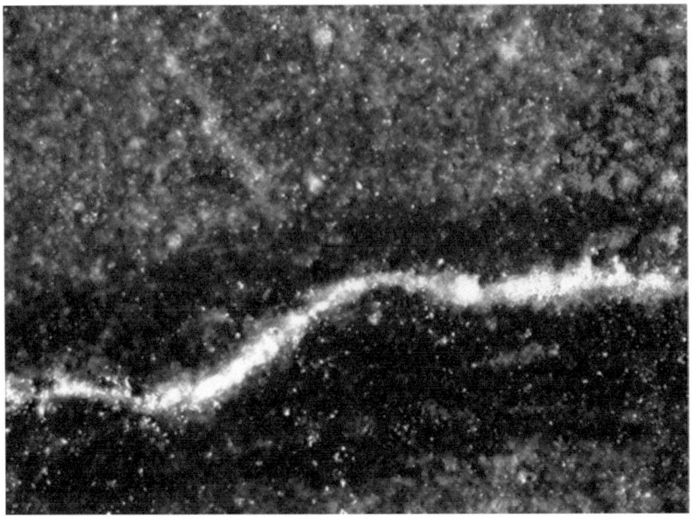

Fig. 4.12 Surface crack closing of a SHCC mix incorporating calcium-sulfoaluminate-based expansive additives (Sisomphon et al. 2013)

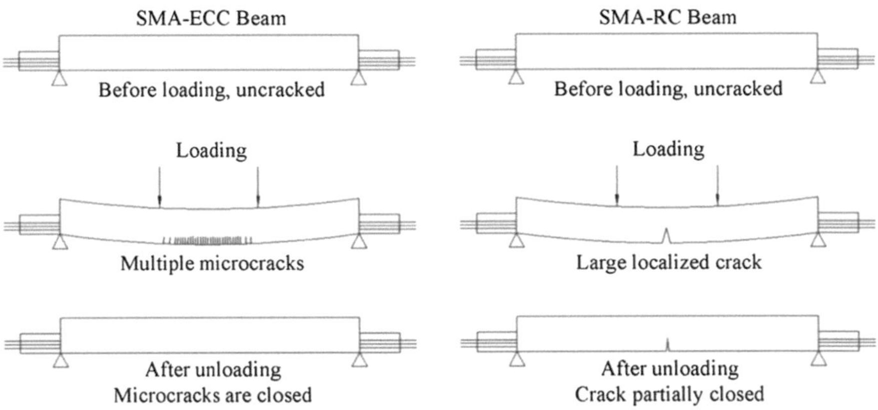

Fig. 4.13 Illustration of the beam behaviour of SMA–ECC, compared with SMA–RC (Li et al. 2015)

4.5 Conclusions

The aging of PVA–SHCC at elevated temperatures and humidity causes an increase in the strength of the matrix and the fibre–matrix interface, and consequently of the composite. The increased strength leads to the reduced ductility of the composite. Although the presence of chlorides in the matrix appears to have little effect at a relatively early age, it does lead to reduced ductility after longer exposure. Leaching reduces the strength characteristics and improves the overall ductility of the composite.

ASR in SHCC has been demonstrated by the inclusion of reactive aggregate and exclusion of fly ash. Experimental evidence of delayed ASR swelling onset and reduced reactivity has been reported for hybrid fibre-reinforced cement-based composites. This is ascribed to the control of ASR swelling-induced cracks by fibre bridging, whereby the ASR product is prevented from escaping from the reaction site. This confinement also causes greater frictional resistance in the fibre–matrix interface, resulting in additional fibre pull-out resistance from an ASR-affected matrix. The pool of experimental results is limited, and the systematic investigation of ASR swelling and deterioration for the various matrix types in terms of the binder content and composition, as well as aggregate content and reactivity, is required to develop appropriate durability design guidelines.

Cracks in SHCC have been shown to heal through autogenic and autonomic processes, provided that the crack widths are small. Autogenic healing occurs in the presence of water and reaction products that can be precipitated. Extra water can be included in fibres or in SAP. Autonomic healing can be achieved by incorporating healing agents in natural or hollow fibres, which are released on fibre rupture. Autonomic crack healing in SHCC has been demonstrated by the addition of bacteria and nutrients, expansive agents and shape-memory alloys.

In order to develop design guidelines that include provision for self-healing, systematic research into matrix type, binder composition and content in SHCC is required. Different fibre types, whether metallic, polymeric, or natural, react differently to chemical attack.

References

Antonopoulou, S. (2009). Self healing in ECC materials with high content of different microfibers and micro particles. MSc Thesis, Delft University of Technology, Delft.

ASTM C-1260 (2007). ASTM, Standard Test Method for Potential Alkali Reactivity of Aggregates (Mortar-Bar Method, D), Annual Book of ASTM Standards, American Society for Testing and Materials.

Beglarigale, A., Yazici, H. (2013). The effect of alkali-silica reaction on steel fiber-matrix bond characteristics of cement based mortars. Construction and Building Materials 47:845-860.

Bektas, F., Turanli, L., Ostertag, C.P. (2006). New approach in mitigating damage caused by alkali-silica reaction. Journal of Material Science 41:5760-5763.

De Rooij, M.R., Qian, S., Liu, H., Gard, W.F., Van de Kuilen, J.W.G. (2009). Using natural wood fibers to self heal concrete. Proceedings of the 2nd International Conference on Concrete Repair, Rehabilitation and Retrofitting (ICCRRR II), 24-26 November 2008, Cape Town, CRC Press.

De Rooij, M., Van Tittelboom, K., De Belie, N., Schlangen, E. (Eds.) (2013). Self-healing phenomena in cement-based materials. Springer, Dordrecht.

Fan, S., Li, M. (2015). X-ray computed microtomography of three-dimensional microcracks and self-healing in engineered cementitious composites. Smart Materials and Structures 24:15-21.

Ferrara, L., Rocha Ferreira, S., Krelani, V., Silva, F., Toledo Filho, R. (2014). Effect of natural fibres on the self-healing capacity of high performance fibre reinforced cementitious

composites. Proceedings of the 3rd International RILEM Conference (SHCC3), 3-5 November 2014, Dordrecht, The Netherlands, RILEM Publications S.A.R.L., pp. 9-16.
Homma, D., Mihashi, H., Nishiwaki, T. (2009). Self-healing capability of fibre reinforced cementitious composites. Journal of Advanced Concrete Technology 7(2):217-228.
Ibrović, V., Luković, M., Schlangen, E. (2014). Decorative application of strain-hardening cementitious composites. Proceedings of the 3rd International RILEM Conference (SHCC3), 3-5 November 2014, Dordrecht, The Netherlands, RILEM Publications S.A.R.L., pp. 381-390.
Kabele, P. (2007). Multiscale framework for modeling of fracture in high performance fiber reinforced cementitious composites. Engineering Fracture Mechanics 74:194-209.
Kabele, P. (2009). Multiscale approach to characterization of damage in fiber reinforced cementitious composites (professor inaugural lecture). CTU in Prague, Prague, Czech Republic.
Kabele, P., Novák, L., Němeček, J., Pekař, J. (2007). Multiscale experimental investigation of deterioration of fiber-cementitious composites in aggressive environment. Proceedings (Jirásek M, Bittnar Z, Mang H eds.) Modeling of Heterogeneous Materials with Application in Construction and Biomedical Engineering, 25-27 June 2007, Prague, Czech Republic, CTU, Prague, pp. 270-271.
Kabele, P., Pekař, J. (2009). Multiscale experimental investigation of PVA-ECC exposed to aggressive environment. Proceedings of the 4th International Conference on Construction Materials ConMat'09: Performance, Innovations and Structural Implications (Uomoto, T., Bantia, N. eds.), JCI, 24-26 August 2009, Nagoya, Japan, pp. 357-362.
Kan, L-L, Shi, H-S. (2012). Investigation of self-healing behavior of Engineered Cementitious Composites (ECC) materials. Construction and Building Materials 29:348-356.
Kim, J.S., Schlangen, E. (2010). Super absorbent polymers to simulate self healing in ECC. Proceedings of the 2nd International Symposium on Service Life Design for Infrastructure, Bagneux, France, RILEM Publications S.A.R.L., pp. 849-858.
Kim, J.S., Schlangen, E. (2011). Self-healing in ECC stimulated by SAP under flexural cyclic load. Proceedings of the 3rd International Conference on Self-healing Materials, 27-29 June 2011, Bath, UK, pp. 324-325.
Li, V.C., Horikoshi, T., Ogawa, A., Torigoe, S., Saito, T. (2004). Micromechanics-based durability study of polyvinyl alcohol-engineered cementitious composite. ACI Materials Journal 101:242-248.
Li, M., Li, V.C. (2011). Cracking and healing of engineered cementitious composites under chloride environment. ACI Materials Journal 108(3):333-340.
Li, M., Sahmaran, M., Li, V.C. (2007). Effect of cracking and healing on durability of engineered cementitious composites under marine environment. Proceedings of the 5th Conference on High Performance Fiber Reinforced Cement Composites (HPFRCC 5), 10-13 July 2007, Mainz, Germany, RILEM Publication S.A.R.L., pp. 313-322.
Li, V.C., Herbert, E.N. (2012). Robust self-healing concrete for sustainable infrastructure. Journal of Advanced Concrete Technology 10:207-218.
Li, V.C., Lim, Y.M., Chan, Y.-W. (1998). Feasibility study of a passive smart self-healing cementitious composite. Composites Part B 29B:819-827.
Li, V.C., Yang, E.H. (2007). Self-healing in concrete materials. Proceedings (Van der Zwaag, S. ed.) Self-healing Materials: An Alternative Approach to 20 Centuries of Materials Science, Springer, Dordrecht, pp. 161-193.
Li, X., Li, M., Song, G. (2015). Energy-dissipating and self-repairing SMA-ECC composite material system. Smart Materials and Structures 24(2):025024.
Nishiwaki, T. Koda, M., Yamada, M., Mihashi, H., Kikuta, T. (2012). Experimental study on self-healing capability of FRCC using different types of synthetic fibers. Journal of Advanced Concrete Technology 10(6):195-206.
Oh, B.H., Kabele, P. (2011). Chapter 3: Durability under chemical loads. In: Durability of strain-hardening cement-based composites (SHCC), State-of-the-art Report of RILEM TC 208-HFC Sub-Committee 2 (Van Zijl & Wittmann eds.), Springer, Dordrecht.

Ostertag, C.P., Blunt, J., Grubb, J. (2007). Mitigation of expansive deterioration processes through crack control. Proceedings of the 6th International Conference on Fracture of Concrete or Concrete Structures (FRAMCOS 6), 17-22 June, Catania, Italy, Taylor & Francis.

Ozbay, E., Sahmaran, M., Yucel, E.H., Erdem, T.K., Lachemi, M., Li, V.C. (2013). Effect of sustained flexural loading on self-healing of engineered cementitious composites. Journal of Advanced Concrete Technology 11:167-179.

Qian, S., Liu, H., Rooij, M.R. de, Schlangen, E., Gard, W.F., Kuilen, J.W.G. van de (2009a). Self-healing cementitious composites for sustainable infrastructures. Proceedings of the 3rd CIB International Conference on Smart and Sustainable Built Environments, June 2009, Delft, pp. 1-9.

Qian, S., Zhou, J., De Rooij, M.R., Schlangen, E., Ye, G., Van Breugel, K. (2009b). Self-healing behavior of strain hardening cementitious composites incorporating local waste materials. Cement and Concrete Composites 31(9):613-621.

Qian, S., Zhang, Z. (2012). Comparison of tensile properties of strain hardening cementitious composite cured in normal and accelerated conditions. Journal of Testing and Evaluation 40(5): 778-783.

Qian, S., Zhou, J., Schlangen, E. (2010). Influence of curing condition and precracking time on the self-healing behavior of engineered cementitious composites. Cement and Concrete Composites 32(9):686-693.

Rokugo, K., Takada, H., Onda, Y., Fujishiro, M., Kobayashi, K., Asano, Y. (2013). Chemical pre-stressing of steel bar-reinforced concrete beams using SHCC that retains ASR expansion for a long time. Proceedings of the 3rd International Conference on Strain Hardening Cement-based Composites (SHCC3), 3-5 November 2014, Dordrecht, The Netherlands, RILEM Publications S.A.R.L., pp. 43-50.

Şahmaran, M., Li, V.C. (2008). Durability of mechanically loaded engineered cementitious composites under alkaline environments. Cement and Concrete Composites 30(2):72-81.

Şahmaran, M., Li, V.C. (2009). Durability properties of micro-cracked ECC containing high volumes fly ash. Cement and Concrete Research 39:1033-1043.

Schlangen, E., Joseph, C. (2009). Self-healing processes in concrete, in self-healing materials: fundamentals, design strategies, and applications, Wiley-VCH, pp. 141-182.

Schlangen, E., Qian, S., Liu, H. (2009). Simulation of self-healing capacity of hybrid fibre material. Proceedings of the 2nd International Conference on Self-Healing Materials, 28 June - 1 July 2009, Chicago, USA.

Sierra Beltran, M.G., Jonkers, H.M., Schlangen, E. (2014a). Characterization of sustainable bio-based mortar for concrete repair. Construction and Building Materials 67:344-352.

Sierra Beltran, M.G., Jonkers, H.M., Schlangen, E. (2014b). Performance of SHCC with bacteria for concrete patch repair. Proceedings of the 15th International Conference and Exhibition on Structural Faults and Repair, 6-8 July 2014, London, The Electrochemical Society, Pennington, pp. 1-18.

Sierra Beltran, M.G., Jonkers, H.M., Schlangen, E. (2015). Self-healing capacity of a strain hardening cement-based composite (SHCC) with bacteria. Proceedings of the 7th RILEM Workshop on High Performance Fibre Reinforced Cement Composites (HPFRCC 7), 1-3 June 2015, Stuttgart, Germany, RILEM Publications S.A.R.L.

Sierra Beltran, M.G., Ukrainczyk, N., Schlangen, E. (2014c). Development of SHCC with biomass fly ash. Proceedings of the 3rd International RILEM Conference on Strain Hardening Cement-based Composites (SHCC3), 3-5 November 2014, Dordrecht, The Netherlands, RILEM Publication S.A.R.L., pp. 87-95.

Sisomphon, K., Copuroglu, O., Koenders, E.A.B. (2011). Self crack healing of strain hardening cementitious composite incorporating expansive agent and crystalline additive. Concrete Solutions, Taylor & Francis, pp. 307-312.

Sisomphon, K., Copuroglu, O., Koenders, E.A.B. (2013). Effect of exposure conditions on self healing behavior of strain hardening cementitious composites incorporating various cementitious materials. Construction and Building Materials 42:217-224.

Snoeck, D., Van Tittelboom, K., Steuperaert, S., Dubruel, P., De Belie, N. (2012). Self-healing cementitious materials by the combination of microfibres and superabsorbent polymers. Journal of Intelligent Material Systems and Structures 25(1):13-24.

Snoeck, D., Smetryns, P.-A., De Belie, N. (2015). Improved multiple cracking and autogenous healing in cementitious materials by means of chemically-treated natural fibres. Biosystems Engineering 139:87-99.

Snoeck, D. (2015). Self-healing and microstructure of cementitious materials with microfibres and superabsorbent polymers. PhD thesis, Ghent University, Ghent, Belgium.

Tziviloglou, E. (2009). Self-healing in ECC materials with low content of different microfibers and micro-particles. MSc thesis, Delft University of Technology, Delft, The Netherlands.

Tziviloglou, E., Van Tittelboom, K., Palin, D., Wang, J., Sierra Beltran, M.G., Ercan, Y.C., Mors, R., Wiktor, V., Jonkers, H.M., Schlangen, E., De Belie, N. (2016). Bio-based self-healing concrete: from research to field application. Advances in Polymer Science 273:345-386.

Van Zijl, G.P.A.G., Wittmann, F.H. (eds.) (2011). Durability of strain-hardening fibre-reinforced cement-based composites (SHCC), State-of-the-Art Report of RILEM Technical Committee 208-HFC, Sub-Committee 2, Volume 4, Springer, Leiden.

Xia Hua (2010). Self-healing of engineered cementitious composites (ECC) in concrete repair system. MSc thesis, Delft University, Delft, The Netherlands.

Yi, C.K., Ostertag, C.P. (2005). Mechanical approach in mitigating alkali-silica reaction. Cement and Concrete Research 35:67-75.

Ying-zi, Y., Lepech, M.D., Li, V.C. (2005). Self-healing of engineered cementitious composites under cyclic wetting and drying. Proceedings of the International Workshop on Durability of Reinforced Concrete under Combined Mechanical and Climatic Loads (CMCL), 27-28 October 2005, Qingdao, China, pp. 231-242.

Chapter 5
Influence of Low Temperatures

Koichi Kobayashi and Folker H. Wittmann

Abstract When strain-hardening cement-based composite (SHCC) with cracks is repeatedly exposed to freezing and thawing, accelerated deterioration of the SHCC, due to the expansion of water during freezing in the cracks, may be of major concern. This chapter summarises the results of studies on the frost damage of SHCC, in particular the frost damage of SHCC with cracks. In most cases described in the literature, the ASTM method C666A — Procedure A (2008) was applied. It has been found that even when cracking has occurred, SHCC has a high resistance with respect to frost damage. When the water-cement ratio is high enough, only a very slight decrease of the relative dynamic modulus of elasticity occurred, although a small amount of scaling was observed on the surface, regardless of the type of fibre or of the composition of the mortar matrix. In tests that simulated the case in which the concrete surface layer damaged by freeze-thaw cycles had been removed and the cross-section had then been repaired with SHCC, it was found that no deterioration of the repaired surface sections occurred due to freezing and thawing. The greater the depth to which the deteriorated concrete was removed, the less susceptible the repaired surfaces were to further freeze-thaw damage.

Keywords Frost damage · Freezing · Freeze-thaw resistance · Dynamic modulus of elasticity

5.1 Introduction

Strain-hardening cement-based composite (SHCC) is a material designed to achieve strain-hardening properties by mixing short high-strength fibres into fine mortar with a maximum aggregate size of typically 0.3 mm. To ensure that the fibres are

K. Kobayashi (✉)
Gifu University, Gifu, Japan
e-mail: ko2ba@gifu-u.ac.jp

F.H. Wittmann
Aedificat Institute Freiburg, Freiburg, Germany

G.P.A.G. van Zijl and V. Slowik (eds.), *A Framework for Durability Design with Strain-Hardening Cement-Based Composites (SHCC)*, RILEM State-of-the-Art Reports 22,
DOI 10.1007/978-94-024-1013-6_5

dispersed uniformly within the mortar, both the fluidity and the viscosity of the mortar matrix must be increased. The fluidity cannot be increased by increasing the water content of the mortar but rather by using a high-performance water-reducing agent. The increase in viscosity is achieved by using large quantities of fine powder and a viscosity-enhancing agent. In this way, a mortar can be obtained that shows no bleeding and which is uniform, having a low water-powder ratio. Moreover, as the cracks that are produced under imposed strain are very fine, SHCC may be expected to have high resistance with respect to frost damage.

5.2 Previous Studies on the Frost Damage Resistance

The results of previous basic studies on the frost damage of SHCC are compiled in one of the International Union of Laboratories and Experts in Construction Materials, Systems, and Structures (RILEM) State-of-the-Art Reports (STARs), edited by Van Zijl and Wittmann (2011): Durability of SHCC (RILEM State-of-the-Art Reports, Vol. 4).

Li et al. (2003) studied the resistance with respect to the rapid repeated freezing and thawing of SHCC using the ASTM C666 (2008) Procedure A (Rapid Freezing and Thawing in Water). Results have shown that, even after 300 cycles of repeated freezing and thawing, no decrease in the dynamic modulus of elasticity occurred in crack-free SHCC samples, and no decrease in tensile toughness was observed.

Şahmaran and Li (2007) induced cracks into SHCC test specimens prepared without air entrainment. Then, they used the ASTM C672 method to conduct tests of slow, repeated freezing and thawing, subsequently studying the scaling resistance involved. Their results have shown that, even when cracking occurred in the SHCC specimens, the amount of scaling was small, and sufficient tensile toughness remained after the conclusion of the tests.

5.3 Effect of Fine Cracks on the Frost Damage Resistance

Liquids can penetrate into cracks instantly by means of capillary suction, if the surface of the material comes into contact with the liquid concerned. According to the water absorption tests conducted by Wittmann et al. (2011), SHCC test specimens with cracks that were induced by means of sustained strain of between 0.5 and 2% absorbed two to three times more water than did crack-free SHCC, when submerged in water for three hours. Such absorption rates were clearly the result of capillary suction. In the tests by Li et al. (2003) described above, the ASTM C666 — Procedure A method was applied to SHCC without induced cracks. However, in the test, the specimens were frozen while they were immersed in water, and, in the case of cracked SHCC, water might possibly freeze inside the cracks, causing volume expansion. Therefore, acceleration of the deterioration could be expected in the case of cracked SHCC.

Yun et al. (2011) applied the ASTM C666 — Procedure A method to perform tests on SHCC specimens, into which cracks had been introduced. Two types of SHCC, one containing PVA fibres, and the other containing PE fibres, were made. From both types, cylindrical test specimens measuring 100 mm in diameter and 200 mm in height were produced and tested under direct tension. Uniaxial tension tests were conducted (a) prior to repeated freezing and thawing, (b) after 100 cycles, and (c) after 200 cycles of repeated freezing and thawing. On the completion of 100 cycles of repeated freezing and thawing, the strain-strain relationship was almost completely unchanged when it was compared to the relationship that was observed prior to the repeated freezing and thawing. After 200 cycles of repeated freezing and thawing, the initial stiffness had increased, whereas the tensile toughness had slightly decreased. The results obtained were explained by means of the fact that the hydration reaction of cement progressed during the freezing and thawing cycles. The observed phenomena were more pronounced with PVA fibres when compared to samples made with PE fibres.

In the investigations described by Rokugo et al. (2008), uniaxial tensile strain of up to 1.0% was imposed, so as to introduce cracks into dumbbell-type test specimens that had been fabricated using PE fibres. Then, freeze-thaw cycles were applied according to the ASTM C666 — Procedure A method, with, finally, a uniaxial tension test being carried out with each of the specimens concerned. Also, in the tests, no deterioration of the mechanical performance of the SHCC due to exposure to the repeated freeze-thaw cycles could be observed, regardless of the existence of cracks.

5.4 Effect of the Mortar Matrix Composition on the Frost Damage

Yun and Rokugo (2012) applied the ASTM C666 — Procedure A method to investigate the frost damage of SHCC fabricated using a mixture of PVA and PE fibres. The fly ash-binder ratio, the silica fume-binder ratio, and the sand-binder ratios were kept constant. They also included Calcium Sulfa Aluminate (CSA) expansive additive at a fixed CSA-binder ratio. Tests were carried out on SHCC with a total of four combinations of the following parameters: two types of water-binder ratio (45 and 55%), and two different volume ratios of the PVA and PE fibres (1.3 and 0.2% in one case, and 0.75 and 0.75% in the other). No cracks were introduced prior to the exposure to the repeated freeze-thaw cycles. For each mix, the decrease in the relative dynamic modulus of elasticity was within 3%, even after 300 freeze-thaw cycles. However, in the case of the mix with 55% water-binder ratio, scaling on the surface of the specimens was observed, with there being a mass loss of approximately 2%. In addition, bending tests were performed after 300 freeze-thaw cycles. In the corresponding test specimens with a volume ratio of PVA and PE of 1.3 and 0.2%, respectively, localised cracking occurred in the SHCC, regardless of the water-binder ratio, and the flexural toughness decreased. Such behaviour was thought to be due to the hydrophilic nature of the PVA, resulting in the deterioration of the interfacial transition zone (ITZ) around the fibres due to freezing and thawing.

Şahmaran et al. (2012) applied the ASTM C666 — Procedure A method to perform freezing and thawing tests, followed by bending tests, on SHCC specimens with two different fly ash-binder ratios. In addition, reference tests were conducted on test specimens fabricated with the same mortar matrix, but without fibres. Despite the fact that no air-entraining agent (AE agent) had been used, the SHCC test specimens had an air content of more than 7%, which meant approximately 1% more than the same mix of the mortar containing no fibres. An examination of the pore size distribution revealed that the SHCC contained many air voids measuring 0.1 mm or greater. The reason was thought to be that the application of hydrophobic oil that had been used to coat the surface of the fibres had resulted in the entrainment of tiny air bubbles. The test specimens made of the mortar matrix containing no fibres were destroyed as a result of repeated freezing and thawing. The SHCC samples, however, exhibited almost no decrease in the relative dynamic modulus of elasticity, and no measurable mass loss, even after exposure to 300 cycles of repeated freezing and thawing. The entrained air bubbles were assumed to reduce the expansion pressure, as were the fibres to prevent crack opening and growth.

5.5 Effect of Repeated Freezing and Thawing on the Bond Between SHCC and Concrete

Structures with frost damage, as well as waterways that have been damaged by abrasion, can be repaired by means of the spraying of a layer of SHCC on the deteriorated surface. In such cases, it is most important that the SHCC applied has sufficient frost resistance, and that the bond between concrete substrate and SHCC overlay does not deteriorate due to repeated freezing and thawing.

Satoh et al. (2014) fabricated prism test specimens measuring 100 × 100 × 200 mm, made out of concrete, with a water-cement ratio of 0.55. Five of their six surfaces were coated with epoxy resin, so that deterioration could occur on only one surface of the test specimens. The specimens were then exposed to 50 freeze-thaw cycles, in accordance with the ASTM C666 — Procedure A method, so as to create surface damage. Subsequently, a water jet was used to remove the deteriorated surface layer, with three different degrees of removal: (1) a depth of approximately 0.5 mm, leading to the exposure of only the fine aggregates (after one water jet application), (2) a depth of approximately 2 mm, with only the top of the coarse aggregates being exposed (after three water jet applications), and (3) a depth of 4 mm, with a considerable section of the coarse aggregate being exposed (after nine water jet applications). Subsequently, the deteriorated surfaces were coated by a layer with a thickness of 10 mm, consisting of three different materials: shrinkage-compensated mortar; polymer cement mortar (PCM); and SHCC. The three types of composite specimen were then exposed to freeze-thaw cycles in accordance with the ASTM C666 — Procedure A method. The results are shown in Fig. 5.1. As the figure shows, the test specimens that were coated with PCM (see Fig. 5.1), and with SHCC (HP in Fig. 5.1) exhibited a reduced decrease of the relative dynamic modulus of elasticity, as compared to the test specimens that were coated with shrinkage-compensated mortar (NS in Fig. 5.1). The PCM- and SHCC-coated specimens exhibited excellent frost damage resistance. In addition, the test specimens for which the removal depth was great (with the numbers 0, 1, 3, and 9 in Fig. 5.1 indicating the number of deteriorating removal operations using the water jet) underwent a relatively small decrease in the relative dynamic modulus of their elasticity and, consequently, a relatively small degree of deterioration during the freeze-thaw cycles.

Fig. 5.1 Influence of the type of repair material on the decrease in the relative dynamic modulus of elasticity under conditions of repeated freezing and thawing (Satoh et al. 2014)

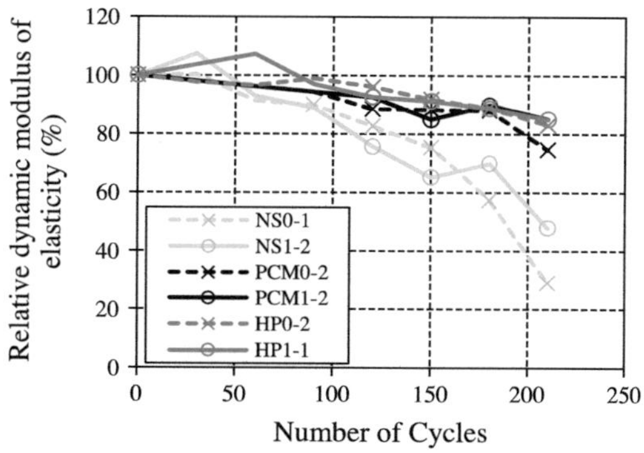

(a) Specimens with shallowly or non-removed substrate.

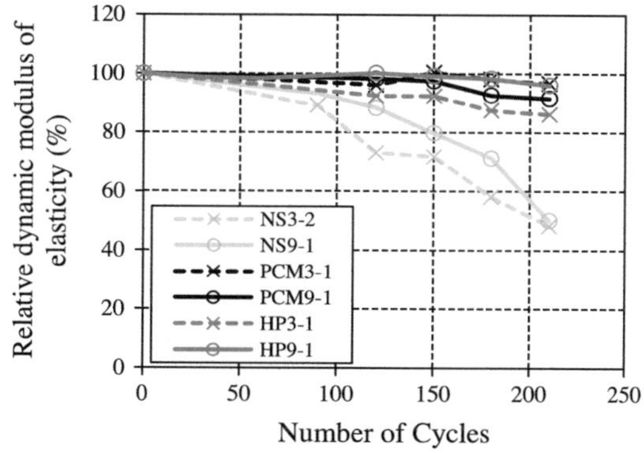

(b) Specimens with normally or deeply removed substrate.

5.6 General Remarks

All results concerning the frost resistance of concrete mentioned in the present section were obtained by applying the ASTM C666 — Procedure A method. RILEM Technical Committee 117-FDC has published a recommendation for the determination of the frost resistance of concrete, using the CDF (capillary suction, de-icing agent, and freeze-thaw) test (1996) and the slab test (2004) as two alternative methods, with both being primarily used for the determination of scaling. More recently, RILEM Technical Committee 176-IDC has published another recommendation, using the CIF (capillary suction, internal damage, and freeze-thaw) test (2004). The CIF test is suggested for the determination of internal damage, together with frost suction, in particular. The RILEM recommendations of the CDF test (1996) and of the slab test (2004) have, in the meantime, become European standards, whereas the CIF test (2004) currently remains a pre-standard. The theoretical basis of all three test methods, which are frequently applied in practice, is described in detail by Setzer (2011).

Damage mechanisms of concrete under freeze-thaw cycles are so complex that they are not, as yet, fully understood in all their detail. In the ASTM C666 — Procedure A method, a thermal gradient is present at four surfaces of the test specimen. In the RILEM test method, however, the thermal gradient is built up almost unidirectionally within the concrete sample. The above reflects the situation in most real structures better than do the conditions that are imposed by ASTM C666 — Procedure A. Therefore, the RILEM test methods are anticipated as being more representative of the damage in real structures than are the other test methods. Hence, it is strongly recommended that comparative tests of SHCC, following the RILEM test methods, are carried out as soon as possible. Wu et al. (2014) report the conducting of freeze-thaw tests on SHCC, according to RILEM methods. They subjected uncracked specimens to 50, 100, 150, and 200 freeze-thaw cycles respectively, after which they performed tensile tests on the specimens concerned, as well as on reference specimens that were not subjected to freeze-thaw cycles. Their results indicate that the ultimate tensile strength and the ultimate tensile strain reduce with an increase in the amount of freeze-thaw cyclic exposure, which, in their case, was by about 15 and 25%, respectively, after 200 cycles had been applied.

5.7 Conclusions

In this chapter, the frost damage of SHCC as induced by freeze-thaw cycles was discussed. The damage of SHCC with cracks under the influence of freeze-thaw cycles and the effect of the composition of the SHCC matrix on frost resistance was dealt with in particular detail. In addition, the behaviour of composite elements consisting of a concrete substrate with an applied surface layer under the influence of freeze-thaw cycles, was discussed. The results of a number of different studies applying the ASTM C666 — Procedure A methodology were presented.

SHCC was shown to have excellent frost damage resistance, mainly because of air entrainment by adding surface-impregnated fibres and because of their crack-bridging capacity, which tends to reduce crack widths. Once cracks form in SHCC, water penetrates quickly into the cracks by capillary absorption. Even then, SHCC shows high resistance to frost damage. When subjected to freeze-thaw cycles according to the ASTM C666 — Procedure A method, the cracked SHCC specimens showed no significant decrease in mechanical properties in either the uniaxial tensile or the bending tests. Although slight scaling was generated when the water–binder ratio in the mortar matrix was high, the SHCC exhibited sufficiently high frost damage resistance for many of the applications.

A further possible reason for the high frost resistance is the unique mixture composition of SHCC, which does not contain coarse aggregates. For this reason, no large ITZ exists in SHCC. In the relatively homogeneous meso- and microstructure of the SHCC, minimum differential thermal stresses between aggregates and mortar matrix occur. These effects require quantitative clarification so as to improve the frost resistance of SHCC even further in future.

Of particular importance is the possibility that microcracks induced by frost action can be closed again due to further hydration of the cement. This process of self-healing requires further investigation, and must be fully understood to optimise the properties of SHCC in general and the resistance to freeze-thaw cycles in particular.

So far, most tests on SHCC have been carried out according to the ASTM C666 — Procedure A method. In CDF tests, however, the test conditions are more severe than in the ASTM tests, making the comparison of results obtained by means of the ASTM method with results obtained in CDF tests of some interest.

When SHCC was applied to repair structural elements that had been damaged by scaling, corresponding test specimens coated with SHCC exhibited high frost damage resistance. Even greater frost damage resistance can be achieved by increasing the depth to which the deteriorated concrete is removed with a water jet prior to administering the coating.

References

ASTM C666 (2008). Standard Test Method for Resistance of Concrete to Rapid Freezing and Thawing. American Society for Testing and Materials.

Li, V.C., Fischer, G., Kim, Y.Y., Lepech, M., Qian, S., Weimann, M., Wang, S. (2003). Durable link slabs for jointless bridge decks based on strain-hardening cementitious components. Report for Michigan Department of Transportation RC-1438, November.

RILEM TC 117-FDC (Freeze-thaw and deicing resistance of concrete) (1996). CDF Test – test method for the freeze-thaw resistance of concrete-tests with sodium chloride solution (CDF). Materials and Structures 29:523-528.

RILEM TC 176-IDC (Internal damage of concrete due to frost action) (2004). Test methods of frost resistance of concrete: CIF - Test: Capillary suction, internal damage and freeze-thaw test – reference method and alternative methods A and B. Materials and Structures 37:743-753.

RILEM TC 176-IDC (Internal damage of concrete due to frost action) (2004). Slab test: freeze/thaw resistance of concrete - internal deterioration. Materials and Structures 37:754-759.

Rokugo, K., Moriyama, M., Kato, H., Lim S.C., Asano Y. (2008). Tensile performance of pre-cracked SHCC after freezing and thawing. Proceedings of 7th RILEM International Symposium on Fiber Reinforced Concrete: Design and Applications - BEFIB2008, 17-19 September 2008, Chennai, India, RILEM Publication S.A.R.L., Bagneux, France, pp. 1071-1078.

Şahmaran, M., Li, V.C. (2007). De-icing salt scaling resistance of mechanically loaded engineered cementitious composites. Cement and Concrete Research 37:1035-1046.

Şahmaran, M, Özbay, E., Yücel, H.E., Lachemi, M., Li, V.C. (2012). Frost resistance and microstructure of engineered cementitious composites: influence of fly ash and micro poly-vinyl-alcohol fiber. Cement and Concrete Composites 34:156-165.

Satoh, A., Shinya, K., Tashiro, K., Rokugo, K. (2014). A proposal on repair methods for freeze-thaw damaged concrete with least re-deterioration. Proceedings of Concrete Solutions 2014, 5th International Conference on Concrete Repair, 1-3 September 2014, Belfast, Northern Ireland, CRC Press, Leiden, Netherlands, pp. 109-114.

Setzer, M.J. (2011). Fundamentals of frost damage in hardened cement paste, in basic research on concrete and applications. Proceedings of an ASMES International Workshop, Wittmann Folker H. and Mercier Olivier (eds.), Aedificatio Publishers Freiburg, Germany, pp. 125-142.

Van Zijl, G.P.A.G., Wittmann, F.H. (eds.) (2011). Durability of strain-hardening fibre-reinforced cement-based composites (SHCC), State-of-the-Art Report of RILEM Technical Committee 208-HFC, Sub-Committee 2, Volume 4, Springer.

Wittmann, F.H., Wang, P., Zhang, P., Zhao, T., Beltzung, F. (2011). Capillary absorption and chloride penetration into neat and water repellent SHCC under imposed strain. Proceedings of the 2nd RILEM Conference on Strain Hardening Cementitious Composites (SHCC2-Rio), 12-14 December, 2011, Rio de Janeiro, Brazil, RILEM Publication S.A.R.L., Bagneux, France, pp. 165-172.

Wu, R.X., Wang, P.G., Wittmann, F.H., Zhao, T.J. (2014). Composition and properties of SHCC; Part II: Influence of elevated temperatures and freeze-thaw cycles on strain hardening cement-based composites. Restoration of Buildings and Monuments 20(2):111-116.

Yun, H.D., Kim, S.W. Lee Y.O., Rokugo K. (2011). Tensile behavior of synthetic fiber-reinforced strain-hardening cement-based composite (SHCC) after freezing and thawing exposure. Cold Regions Science and Technology 67:49-57.

Yun, H.D., Rokugo, K. (2012). Freeze-thaw influence on the flexural properties of ductile fiber-reinforced cementitious composites (DFRCCs) for durable infrastructures. Cold Regions Science and Technology 78:82-88.

Chapter 6
Influence of Elevated Temperatures

Flavio A. Silva, Barzin Mobasher, Alva Peled, Dimas A.S. Rambo and Romildo D. Toledo Filho

Abstract In this chapter, the effect of elevated temperatures on strain-hardening cement-based composites (SHCC) is reported. Key features of SHCC such as tensile strength and strain capacity, compressive strength, failure modes, the fiber–matrix interface, and spalling behavior are discussed. Different testing conditions are covered, including not only residual but also high temperature tests. Experimental results addressing various temperature levels (ranging from ambient temperature to 1000 °C) were used to investigate basic knowledge of the thermo-mechanical response of SHCC under conditions of both uniaxial tension and compression.

Keywords Thermal effects · Elevated temperature · SHCC · PVA · Fabrics

6.1 Introduction

Two main aspects of the thermal effects of strain-hardening cement-based composite (SHCC) materials deal with the effect of temperature on the fibers and matrix phases. The effects can be categorized into long-term and short-term effects, and may involve thermally assisted chemical degradation, conversion reactions and physical effects. Concrete structures such as tunnels and buildings may be exposed to fires and other conditions that can cause surface temperatures of more than 1000 °C. In such cases, differential temperature rise plays an important role in the

F.A. Silva (✉)
Pontifícia Universidade Católica do Rio de Janeiro, Rio de Janeiro, Brazil
e-mail: fsilva@puc-rio.br

B. Mobasher
Arizona State University, Tempe, USA

A. Peled
Ben-Gurion University of the Negev, Beersheba, Israel

D.A.S. Rambo · R.D. Toledo Filho
Universidade Federal do Rio de Janeiro, Rio de Janeiro, Brazil

G.P.A.G. van Zijl and V. Slowik (eds.), *A Framework for Durability Design with Strain-Hardening Cement-Based Composites (SHCC)*, RILEM State-of-the-Art Reports 22,
DOI 10.1007/978-94-024-1013-6_6

ensuing mechanical thermal degradation, while heat transfer can raise the interior of the concrete to any temperature in the range 300–700 °C; during prolonged periods of mechanical thermal degradation, chemical decomposition becomes a major concern (Kim et al. 2015). The effect of sustained high temperatures in Portland-cement-based concrete materials leading to a degradation of the material properties (e.g. strength and modulus of elasticity) has been studied in various forms by different researchers (Balázs and Lublóy 2012; Chan et al. 2000; Handoo et al. 2002; Komonen and Penttala 2003; Li et al. 2004; Naus 2006). One aspect of this effect is the physical–chemical changes of the cement paste and aggregates, the pore structure, and the thermal incompatibility among different constituents that can cause internal microcracking. The mechanical properties of concrete greatly depend on the hydration products (calcium silicate hydrate gel, calcium hydroxide, and ettringite) contained therein. A temperature above 350 °C can cause the loss of chemically bonded water and the decomposition of calcium hydroxide, as well as an increase in the pore size and porosity of the hydrated matrix (Çavdar 2012). The other form of damage is the spalling due to the internal tensile stress induced by the vapor pressure that occurs under conditions of rapid heating in the temperature range of 200–350 °C (Komonen and Penttala 2003).

The behavior of SHCC at elevated temperature is of great importance for the safe operation of structural components and systems. In certain cases the temperature may routinely oscillate between 60 and 260 °C, and, under accident conditions, temperatures may reach or exceed 600 °C (Willam et al. 2009). Under these conditions, fiber-reinforced concrete (FRC) materials can experience a severe reduction in mechanical properties such as compressive strength, with the critical temperature being around 400 °C (Balázs and Lublóy 2012; Çavdar 2012, 2013). By 650 °C, FRC containing various fibers such as polypropylene (PP), polyvinyl alcohol (PVA), carbon, glass, and aramid fibers has shown a 60–70% decrease in its compressive strength, with major cracks becoming visible (Khaliq and Kodur 2012).

6.2 Degradation of PVA–SHCC Under Elevated Temperatures

Bhat et al. (2014) studied the residual effect of temperature on PVA-reinforced strain-hardening cement composites up to 600 °C, resulting in observable deterioration in tensile strength and strain capacity. At 200 °C, the degradation in the fiber/matrix interfacial properties led to a strong (more than 50%) reduction in the tensile strain capacity, whereas the tensile strength increased slightly, by around 1%. At temperatures above 200 °C, the tensile strain-hardening behavior diminished, whereas the tensile strength was reduced by approximately 40%. No spalling was observed even after 6 h of constant exposure to 600 °C, which was attributed to the tensile strain capacity, as well as to the increase in porosity due to fiber melting. The formation of pores caused a decrease in the compressive strength involved, but it allowed water vapors to escape, lessening the vapor pressure and tensile stress concerned (Chen and Liu 2004; Sideris et al. 2009).

Magalhães et al. (2009, 2015) studied the residual tensile strength of PVA–SHCC after exposure to temperatures of 90–250 °C (in the residual condition). The loss of the composite mechanical performance was observed in terms of reduced stiffness, tensile strength, and ductility, as well as in changes in the cracking pattern. For the specimens that were preheated at 250 °C, the strain capacity presented 92% reduction (from 2.98 to 0.24%), whereas the tensile strength decreased by 68% (from 2.90 to 0.92 MPa).

Mechtcherine et al. (2012) performed both residual and hot-stage tensile tests on PVA–SHCC in temperatures ranging from 22 to 150 °C, and at various strain rates, with the latter being more severe than were the former (see Fig. 6.1). When testing SHCC with a hot-stage apparatus, a reduction in the first-crack strength was observed. At 60 and 100 °C, a reduction of the bond strength between the fiber and matrix was observed when the composite was tested in combination with the elevated temperatures. As a result of the weaker bond, longer fiber-free lengths could develop and higher elastic and plastic deformation of fibers occurred, leading to an increase in the strain capacity of the composite concerned. At 150 °C, the SHCC lost its ductility and the ability to form multiple cracks, due to a strong decrease in the fibers' strength (see Fig. 6.2).

Oliveira et al. (2014), Oliveira (2015) performed direct tensile tests on PVA–SHCC subjected to elevated temperatures (hot-stage tests at 22, 60 and 90 °C), and at different levels of the internal moisture conditions (95, 50, 20 and 0%), and to combinations of these parameters. The results indicate that the SHCC tensile strength decreased, and that the strain capacity increased with an increase in temperature (see

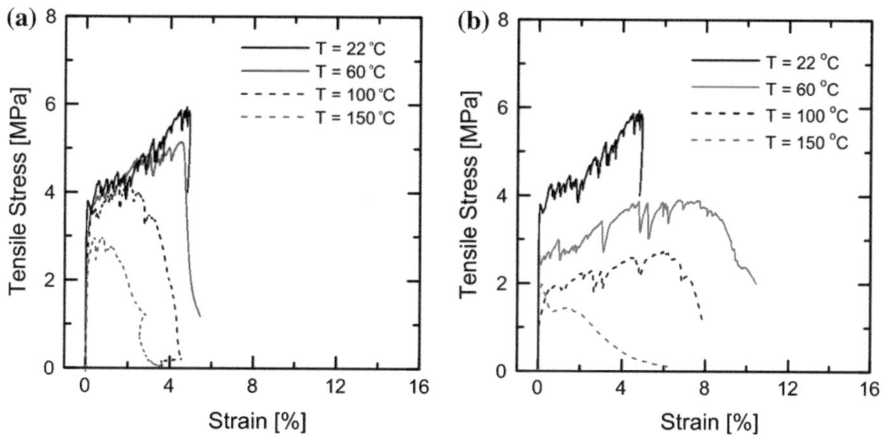

Fig. 6.1 Effect of temperature on the **a** residual and **b** hot-stage tensile stress–strain response of SHCC, tested at the strain rate of 3.10^{-4} s^{-1} (Mechtcherine et al. 2012)

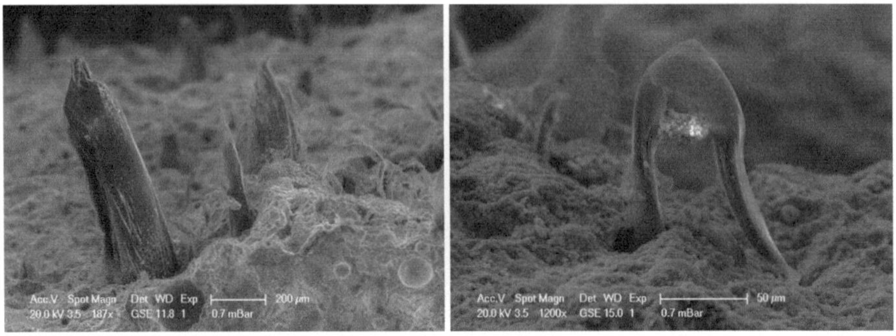

Fig. 6.2 Deterioration of PVA fibers observed inside the SHCC, after tensile tests performed at 150 °C (Mechtcherine et al. 2012)

Fig. 6.3). Such change could be traced back to a softening of the fiber coating that resulted in a modification of the fiber–matrix interface properties concerned. Furthermore, in general, by reducing the internal humidity of the specimens (from 95 to 0%), the strain capacity involved was reduced at elevated temperatures.

The decrease in the first-crack strength that was noticed by Oliveira et al. (2014), Oliveira (2015) can be explained as follows. The cement paste, aggregates and PVA fiber have different Poisson's ratio. Thus, as the fiber has a higher Poisson's ratio, it deforms transversely more than do the other elements of SHCC when they are subjected to the same strains. Mostly, the Poisson's ratio of the materials can be even more distinct when the materials are exposed to moderate temperatures (60 and 100 °C). Furthermore, the SHCC's thermal expansion is non-uniform. The thermal expansion coefficient α for the SHCC is reported in the literature as being 10.35 $\mu\varepsilon$ °C^{-1}. However, the materials composing the SHCC have distinct values of α. For sand, the value is 11.8 $\mu\varepsilon$ °C^{-1}, whereas, for cement paste, the value varies between 14 and 20 $\mu\varepsilon$ °C^{-1}, and for the PVA fiber the value is 4.4 $\mu\varepsilon$ °C^{-1}. When a SHCC specimen is exposed to in situ high temperature, it is probable that it can present differences in its expansion in terms of the cement paste, the aggregate, and the PVA fiber. Under such circumstances, changes in first-crack strength can be justified.

In Oliveira's (2015) pull-out tests on hot-stage condition, it was observed that the PVA fiber–matrix interface was quite sensitive to elevated temperatures. With a temperature increase from 22 to 100 °C, the maximum force, the chemical nominal bond, and the friction bond were reduced (see Fig. 6.4). Such behavior was noted as also being in agreement with the results that were presented in the research of Mechtcherine et al. (2012). When SHCC that was reinforced by means of PVA fibers was tested at moderate temperatures (up to 100 °C), a drop was experienced in the strength, and an increase in strain that, at least in part, could be traced back to a decrease in the fiber–matrix bond strength.

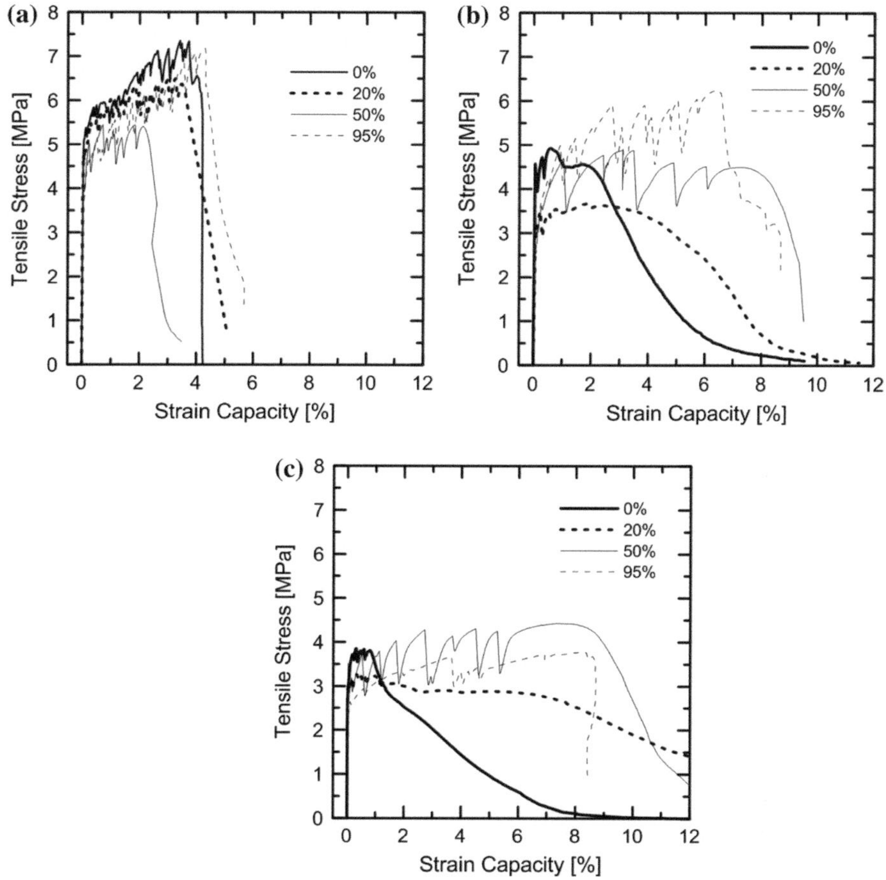

Fig. 6.3 The combined influence of temperature and humidity on the tensile behavior of PVA–SHCC at **a** 22 °C, **b** 60 °C, and **c** 90 °C (Oliveira 2015)

Yu et al. (2015) investigated the mechanical behavior of high-volume fly ash SHCC after exposure of the composites to temperatures ranging from 20 to 200 °C. The tensile strength and strain capacity increased after exposure to temperatures of 50 and 100 °C, but decreased after exposure to a temperature of 200 °C.

Studies developed by Wu et al. (2014), with three different types of SHCC, showed that the maximum crack width of the observed crack width distribution increases after exposure to elevated temperatures. This behavior causes vulnerability in the composite performance in aggressive environments. The authors also performed tests with samples that were exposed to freeze–thaw cycles. The maximum number (200) of freeze–thaw cycles that were used in the study had almost no influence on the stress–strain diagram of SHCC.

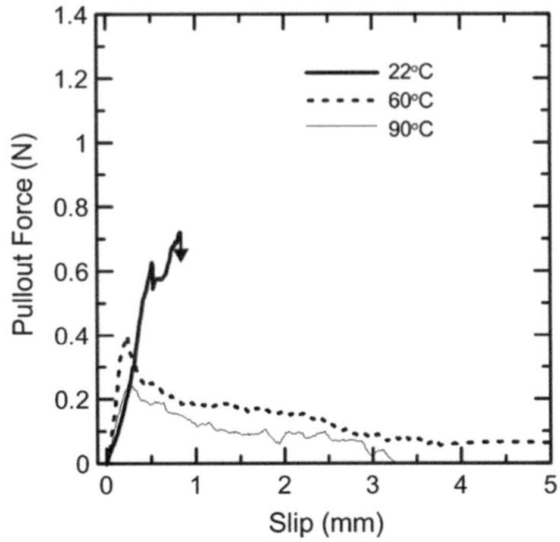

Fig. 6.4 Effect of moderate temperatures on the PVA fiber–matrix interface (Oliveira 2015)

Chen and Liu (2004) showed that the residual compressive strength of high-strength concrete starts to decrease above 200 °C, yielding only 10% of its original strength at 800 °C. Çavdar (2012) demonstrated that carbon fibers, PP, glass, and PVA are active in cement-based matrices under bending and compressive loads only up to a temperature of 450 °C.

6.3 Degradation of SHCC Reinforced with Other Types of Fibers

When a coating is used in SHCC, the bond performance between fibers and the cementitious matrix can change with the temperature. Silva et al. (2014), using carbon fibers as reinforcement, showed that, when heating polymer-coated carbon fibers up to a temperature of 150 °C, a polymer-interlocking mechanism between filaments and matrix was observed. The mechanism resulted in significant increases in the maximum pull-out load. Krüger and Reinhardt (2006) performed fire tests on four different I-shaped mortar beams that were reinforced with AR glass and carbon fibers. The investigation focused on the load-bearing capacity of the composite during a fire test under constant load. In one of the cases, an SBR (styrene–butadiene) thermoplastic resin was used as coating of the carbon fiber, with the results concerned being found to be dependent on the fire behavior of the fibers involved. Due to the softening of the SBR coating (at a temperature of about 90 ° C), the fiber–matrix interface was rapidly impaired, resulting in fiber pull-out, and subsequent failure.

Rambo et al. (2015) investigated the residual thermo-mechanical properties of a refractory composite that was reinforced by means of polymer-coated basalt fiber under tensile loading at varying temperatures, ranging from 25 to 1000 °C. The matrix–polymer interlocking mechanism was found to help create a high bond between the filaments and the matrix concerned in those instances in which the sample was preheated in a temperature up to 150 °C. This processing step was found to have a significant impact on the maximum tensile response of the composite. Furthermore, yarn strength was dependent on the coating penetration, which, after cooling, promoted load redistribution between the filaments present. After preheating the composite to a temperature above 150 °C, a clear drop in the tensile performance was observed, due to the thermal decomposition of the coating, and due to the dehydration process of the matrix. The behavior of the composite became brittle after exposure to temperatures of 600 and 1000 °C, reducing the tensile strength to values lower than those at room temperature. Rambo (2016) also investigated the mechanical behavior of the same refractory composite submitted to high-temperature direct tensile tests up to 400 °C. The results showed that a much more aggressive loss of load-carrying capacity was observed under hot conditions. Values for the ultimate tensile strength obtained under such conditions were, on average, about 50% lower than the residual ones. According to the author, the detrimental effect was, at least in part, attributable to the viscoelastic/plastic changes of the polymer with the varying temperature, as well as to the high pore pressures that, when combined with thermal cracking, can substantially reduce the mechanical performance of the composite under hot conditions.

Yao et al. (2016) investigated the interaction of high strain rates and varying temperature on cementitious systems that were reinforced with glass and PP fibers. The high-speed tensile tests were performed under a nominal strain rate of 100 s^{-1}, while the temperature ranged from -30 to 80 °C. Both the tensile strength and the post-cracking stiffness in various composites decreased with increasing temperature. A tension-stiffening model that was developed by Soranakom and Mobasher (2010a, b) was used to simulate the tensile response of textile-reinforced concrete (TRC). As shown in Fig. 6.5c, d, the degradation in post-cracking stiffness can be simulated by the model using decreased efficiency of the yarn stiffness $\eta < 1$, as well as the decreased fiber–matrix interfacial bond strength, respectively. The efficiency factor is defined to represent the inefficiencies in the bonding that lead to telescopic, or sleeve, effect and that has been quantified by means of experiments on sleeve filaments that bond to matrix, and that contribute to axial stiffness, whereas the core filaments provide marginal stiffness, due to the presence of unbonded yarns (Banholzer et al. 2006; Cohen and Peled 2010). The simulation indicates that, as the temperature increases, the interface bond strength decreases, with the telescopic effect in the yarn bundles becoming increasingly pronounced (Fig. 6.5d).

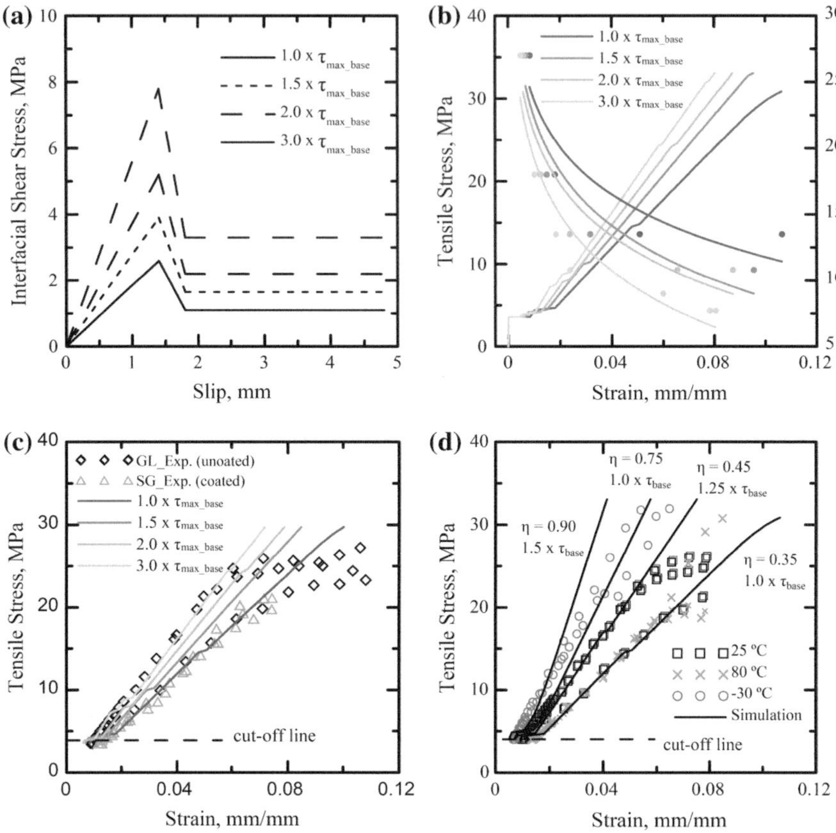

Fig. 6.5 **a** Interfacial bond model used in tension stiffening; **b** effect of bond strength on post-crack tensile responses and average crack spacing; **c** experimental and numerical tensile stress–strain response of GL and SG-TRCs at room temperature; **d** experimental and numerical tensile stress–strain response of GL composites at different temperatures Soranakom and Mobasher (2010a, b)

6.4 Conclusions

This chapter discussed the mechanisms responsible for the loss of load-bearing capacity of SHCC under tensile/compression load and elevated temperatures. Noticeably, when PVA–SHCC is tested under hot conditions, the degradation is more severe than in the case of residual tests. For hot-stage tests, the first crack and ultimate tensile strength tend to decrease, whereas the strain capacity increases for temperatures up to 100 °C. Above 150 °C, the PVA fibers lose their strength and ductility. In the residual tests the SHCC kept its first-crack strength up to 100 °C, but reduction in both strain and strength capacity was observed when increasing the temperature above 100 °C.

For SHCC reinforced with polymer-coated fibers, the interface plays a major role. A matrix–polymer interlocking mechanism takes place at temperatures up to 150 °C to create high bonding between filaments and matrix when the SHCC is preheated. A drop in the tensile performance was observed due to the thermal decomposition of the coating between 150 and 200 °C and the dehydration process of the matrix. For hot-stage direct tensile tests, a much more aggressive loss of load-carrying capacity was observed. For carbon and basalt fabrics, exposure to a temperature higher than 400 °C results in major degradation.

References

Balázs, G.L., Lublóy, E. (2012). Post-heating strength of fiber-reinforced concretes. Fire Safety Journal 49:100-106.

Banholzer, B., Brockmann, T., Brameshuber, W. (2006). Material and bonding characteristics for dimensioning and modeling of textile reinforced concrete (TRC) elements. Materials and Structures 39:749-763.

Bhat, P.S., Chang, V., Li, M. (2014). Effect of elevated temperature on strain-hardening engineered cementitious composites. Construction and Building Materials 69:370-380.

Çavdar, A. (2012). A study on the effects of high temperature on mechanical properties of fiber reinforced cementitious composites. Composites Part B: Engineering 43(5):2452-2463.

Çavdar, A. (2013). The effects of high temperature on mechanical properties of cementitious composites reinforced with polymeric fibers. Composites Part B: Engineering 45(1):78-88.

Chan, Y.N., Luo, X., Sun, W. (2000). Compressive strength and pore structure of high performance concrete after exposure to high temperature up to 800°C. Cement and Concrete Research 30(2):247-251.

Chen, B., Liu, J. (2004). Residual strength of hybrid-fiber-reinforced high-strength concrete after exposure to high temperatures. Cement and Concrete Research 34:1065-1069.

Cohen, Z., Peled, A. (2010). Controlled telescopic reinforcement system of fabric–cement composites — durability concerns. Cement and Concrete Research 40:1495-1506.

Handoo, S.K., Agarwal, S., Agarwal, S.K. (2002). Physicochemical, mineralogical, and morphological characteristics of concrete exposed to elevated temperatures. Cement and Concrete Research 32:1009-1018.

Khaliq, W., Kodur, V. (2012). High temperature mechanical properties of high-strength fly ash concrete with and without fibers. ACI Materials Journal 109:665-674.

Kim, J., Lee, G.P., Moon, D.Y. (2015). Evaluation of mechanical properties of steel-fiber-reinforced concrete exposed to high temperatures by double-punch test. Construction and Building Materials 79:182-191.

Komonen, J., Penttala, V. (2003). Effects of high temperature on the pore structure and strength of plain and polypropylene fiber reinforced cement pastes. Fire Technology 39:23-34.

Krüger, M., Reinhardt, H.W. (2006). Composite materials – fire resistance. W. Brameshuber (ed.), Report 36: Textile Reinforced Concrete – State-of-the-Art Report of RILEM Technical Committee 201-TRC, Chapter 6, RILEM Publications S.A.R.L, Bagneux, France, pp. 211-219.

Li, M., Qian, C., Sun, W. (2004). Mechanical properties of high-strength concrete after fire. Cement and Concrete Research 34:1001-1005.

Magalhães, M.S., Toledo Filho, R.D., Fairbairn, E.M.R. (2015). Thermal stability of PVA fiber strain hardening cement-based composites. Construction and Building Materials 94:437-447.

Magalhães, M.S., Toledo Filho, R.D., Fairbairn, E.M.R. (2009). Physical and mechanical properties of strain-hardening cement-based composites (SHCC) after exposure to elevated temperatures. G.P.A.G. van Zijl, W.P. Boshoff (eds.). Proceedings of the International

Conference on Advanced Concrete Materials (ACM), 17-19 November 2009, Stellenbosch, South Africa, CRC Press, Leiden, The Netherlands, 2010, pp. 203-207.
Mechtcherine, V., Silva, F.D.A., Müller, S., Jun, P., Toledo Filho, R.D. (2012). Coupled strain rate and temperature effects on the tensile behavior of strain-hardening cement-based composites (SHCC) with PVA fibers. Cement and Concrete Research 42(11):1417-1427.
Naus, D. (2006). The effect of elevated temperature on concrete materials and structures – a literature review. Technical Report, Oak Ridge National Laboratory, March 2006.
Oliveira, A.M., Silva, F.A., Fairbairn, E.M.R., Toledo Filho, R.D. (2014). Temperature and internal moisture effects on the tensile behavior of strain-hardening cement-based composites (SHCC) reinforced with PVA fibers. E. Schlangen, M.G. Sierra Beltran, M. Luković, G. Ye (eds.), Proceedings of the 3rd International RILEM Conference on Strain Hardening Cementitious Composites, 3-5 November 2014, Dordrecht, The Netherlands, RILEM Publications S.A.R.L., Bagneux, France, pp. 51-60.
Oliveira, A.M. (2015). On the behavior of PVA SHCC under the combined temperature and humid effects and sustained loads. PhD thesis, Department of Civil Engineering, COPPE/UFRJ, Brazil.
Rambo, D.A.S., Silva, F.A., Toledo Filho, R.D., Gomes, O.F.M. (2015). Effect of elevated temperatures on the mechanical behavior of basalt textile reinforced refractory concrete. Materials and Design 65:24-33.
Rambo, D.A.S. (2016). Temperature effect on the mechanical behavior of textile reinforced refractory composites. PhD thesis, Department of Civil Engineering, COPPE/UFRJ, Brazil.
Sideris, K.K., Manita, P., Chaniotakis, E. (2009). Performance of thermally damaged fiber reinforced concretes. Construction and Building Materials 23:1232-1239.
Silva, F.D.A., Butler, M., Hempel, S., Toledo Filho, R.D., Mechtcherine, V. (2014). Effects of elevated temperatures on the interface properties of carbon textile-reinforced concrete. Cement and Concrete Composites 48:26-34.
Soranakom, C., Mobasher, B. (2010a). Modeling of tension stiffening in reinforced cement composites: Part I. Theoretical modeling. Materials and Structures 43:1217-1230.
Soranakom, C., Mobasher, B. (2010b). Modeling of tension stiffening in reinforced cement composites: Part II. Simulations versus experimental results. Materials and Structures 43:1231-1243.
Willam, K., Xi, Y., Lee, K., Kim, B. (2009). Thermal response of reinforced concrete structures in nuclear power plants. SESM No. 02-2009, Research Report for Oak Ridge National Laboratory.
Wu, R., Wittmann, F.H., Wang, P., Zhao, T. (2014). Influence of elevated and low temperature on properties of SHCC. E. Schlangen, M.G. Sierra Beltran, M. Luković, G. Ye (eds.), Proceedings of the 3rd International RILEM Conference on Strain Hardening Cementitious Composites, 3-5 November 2014, Dordrecht, The Netherlands, RILEM Publications S.A.R.L., Bagneux, France, pp. 3-8.
Yao, Y., Bonakdar, A., Faber, J., Gries, T., Mobasher, B. (2016). Distributed cracking mechanisms in textile-reinforced concrete under high speed tensile tests. Materials and Structures 49:2781-2798.
Yu, J., Lin, J., Zhang, Z., Li, V. (2015). Mechanical performance of ECC with high-volume fly ash after sub-elevated temperatures. Construction and Building Materials 99:82-89.

Chapter 7
Abrasion

Volker Slowik, Steffen Müller, Christian Wagner and Viktor Mechtcherine

Abstract The abrasion resistance of strain-hardening cement-based composite (SHCC) surfaces has been investigated experimentally by different methods. The results obtained are difficult to compare. Using the Aggregate Wear Index (AWI), commonly used for measuring abrasion resistance in the USA, it can be demonstrated that SHCC may have sufficient abrasion resistance to cope with heavy road traffic. Experiments conducted with the *Böhme* grinding wheel revealed a relationship between abrasion resistance and the tensile strength of the material. The friction coefficient of SHCC surfaces was also measured. After appropriate surface texturing, the friction coefficient may be sufficient for roads that are subjected to heavy traffic.

Keywords Wear · Friction · Grinding wheel · Texturing

7.1 Introduction

Under some service conditions, strain-hardening cement-based composite (SHCC) surfaces may be subjected to abrasion, in particular when they are used for the patch repair of concrete pavements or for road overlays. Accordingly, the abrasion resistance has to be considered in the durability design.

High abrasion resistance of cement-based materials is usually achieved by coarse aggregates, i.e., by means of a comparatively large maximum aggregate size and a high aggregate content. SHCC does not have these characteristics, which makes achieving sufficient abrasion resistance a technically challenging task. Another strategy for improving abrasion resistance entails the dry-shake application of hard aggregate to the surface. Experience has shown, however, that these hard particles

V. Slowik (✉) · C. Wagner
Leipzig University of Applied Sciences, Leipzig, Germany
e-mail: volker.slowik@htwk-leipzig.de

S. Müller · V. Mechtcherine
Technische Universität Dresden, Dresden, Germany

G.P.A.G. van Zijl and V. Slowik (eds.), *A Framework for Durability Design with Strain-Hardening Cement-Based Composites (SHCC)*, RILEM State-of-the-Art Reports 22,
DOI 10.1007/978-94-024-1013-6_7

may be extracted from SHCC surfaces when subjected to grinding. Under this action, the matrix of the SHCC may not be strong enough to retain these particles. When the particles are separated from the surface, they act as additional grinding media and may even reduce the wear resistance. Another problem stems from the fact that the test standards for quantifying abrasion resistance vary so significantly that it is difficult to compare the test results obtained according to the different standards in existence.

Recently, so-called hydro-abrasion has also been the subject of investigations, in particular in Germany and Japan. At the TU Dresden, hydro-abrasion experiments were performed within the framework of a Master's thesis (Rohde 2012), using a rotating drumlike container filled with water and steel balls. The combination of sliding and impact wear caused a more pronounced deterioration of SHCC in comparison to what tends to occur in ordinary concrete. In Japan, the interest in hydro-abrasion results from the need to ensure the durability of SHCC used for agricultural waterways. The results of the corresponding study have not been published yet.

7.2 Abrasion Tests and Discussion of the Results Obtained

Li and Lepech (2004) were the first to publish abrasion test results for SHCC. They conducted static friction and wear testing, according to the Michigan Test Method 111 (MDOT 2001). After determining the initial static friction forces between a textured wet surface and a test tyre, the surface was subjected to four million tyre passes. The friction forces after the exposure yield the so-called Aggregate Wear Index (AWI), with them being determined so as to evaluate the effect of surface polishing or deterioration. The measured AWI values ranged between 1.6 and 2.3 kN, which exceeded the value of 1.2 kN that is required for Michigan trunk-line road surfaces. The material under investigation was concluded to have sufficient abrasion and wear resistance to heavy traffic. This conclusion was confirmed by means of field experience with the patch repair of a Michigan bridge deck that was carried out in 2002, and which was monitored throughout 2007 (Li and Li 2008). No significant abrasion has yet been reported in connection with the deck.

Especially in European countries, a completely different method is used for quantifying the abrasion resistance of cementitious materials. This method is based on the application of the so-called *Böhme* grinding wheel, which is standardised in DIN 52108:2010-05 (2010). The test apparatus consists of a rotating disc, onto which the specimen surface is pressed. Reproducible abrasion conditions are ensured by means of the use of a standardised grinding medium. Wagner (2007) performed tests on SHCC surfaces, according to the 2006 version of the standard. The results and the material composition were published in a RILEM (International Union of Laboratories and Experts in Construction Materials, Systems and Structures) State-of-the-Art Report (Van Zijl and Wittmann 2011). After 28 days, and based on the mass loss experienced, an abrasion of 16.9 $cm^3/50\ cm^2$ was measured (see Table 7.1). According to the result obtained, the SHCC used in the tests concerned could only be assigned to the lowest abrasion class, according to the stipulations set out in DIN EN 13813:2003-01 (2003), which would not be appropriate for heavy roadway traffic. However, the abrasion of the SHCC was 23% lower than was the abrasion of a reference mortar without fibres.

Similar tests with the *Böhme* grinding wheel were performed in 2012, by the Institute of Construction Materials at the TU Dresden (Wellner et al. 2016). A significantly higher abrasion resistance than that which is mentioned above was measured, even for an SHCC surface that was predamaged by means of exposure to freeze–thaw cycles. Table 7.1 contains the results that were obtained in the tests. Based on these data, the undamaged SHCC tested in the present instance can be assigned to abrasion class A12, according to DIN EN 13813:2003-01 (2003). Noteworthy, the ratio between the abrasion of the SHCC and the fibreless reference mortar was found to be the same as in the case of Wagner's (2007) results, with the age of testing being 28 days in both cases. Table 7.2 enumerates the composition of the SHCC that was used in Dresden. The major differences to the material, as tested by Wagner (2007), were the higher cement strength and the higher cement content, whereas the water–binder ratio was the same in both cases. This resulted in a comparatively high tensile strength of 5.0 MPa for the undamaged, and 4.2 MPa for the predamaged SHCC, as tested at the TU Dresden. The tensile strength of the material that was tested by Wagner (2007) was 2.81 MPa, with a standard deviation of 0.17 MPa. In addition, Wagner (2007) used 8 mm polyvinyl alcohol (PVA) fibres, whereas the PVA fibre length in the experiments at the TU Dresden

Table 7.1 Abrasion determined according to *Böhme* (DIN 52108)

	Tensile strength (MPa)	Abrasion based on mass loss ($cm^3/50\ cm^2$)		
		Reference mortar without fibres	SHCC	SHCC after 28 freeze–thaw cycles
Wagner (2007)	2.81	21.90	16.90	–
Wellner et al. (2016)	5.00	14.87	11.59	–
	4.20 (after 28 freeze–thaw cycles)	–	–	13.61

Table 7.2 Material composition for Wellner et al.'s (2016) abrasion tests

Component	kg/m^3
CEM I 42.5 R-HS	505
Fly ash	621
Fine sand (0.06 … 0.2 mm)	536
Stabiliser	3.2
Plasticiser	10
Water	338
PVA Fibre Kuralon K-II (12 mm)	26

was 12 mm. It has also to be considered that Wagner (2007) dried the samples at 105 °C prior to the abrasion testing, whereas the samples in Dresden were tested directly after storage at 20 °C and at 65% relative humidity.

The improved abrasion resistance, with respect to Wagner's (2007) study, is mainly attributed to the relatively high strength of the matrix. This conclusion is supported by the 2014 test results that were also obtained by the group at the TU Dresden. The same SHCC composition was used as in 2012, however a cement of higher strength and a different type of fly ash was employed. The composition concerned resulted in an increase in the SHCC tensile strength, from 5.0 MPa in 2012 to 5.5 MPa in 2014. The material loss due to abrasion decreased from 11.59 $cm^3/50\ cm^2$ in 2012 to 4.31 $cm^3/50\ cm^2$ in 2014. Thus, there appears to be a relationship between the abrasion resistance according to *Böhme* and the tensile strength of the material.

7.3 Friction Coefficient of Surfaces

Another test series that was conducted by the Institute of Construction Materials at the TU Dresden was aimed at evaluating the friction coefficient of differently textured SHCC surfaces. Specimens with a surface area of 38 cm by 48 cm, and with a height of 12 cm, were prepared. The surface finishing was done by (a) regular smoothing, (b) intensified smoothing, (c) texturing by means of dragging burlap over the surface, (d) texturing by means of using a steel broom, and (e) oriented texturing by means of grinding the surface. The last-mentioned method is similar to grooving, but with the grooves involved being of considerably higher density. Texturing by means of using burlap appeared not to be applicable for SHCC surfaces, with its effect tending to be either irregular or negligible. After the surface texturing, drill cores with a diameter of 220 mm were extracted from the specimens and tested by the *Wehner/Schulze* method, which enables traffic simulation by means of consecutive polishing periods and friction measurements. First, the friction coefficient of the original surface is measured. The friction measurement is then repeated after 90,000 and 180,000 polishing cycles, respectively. After 90,000 cycles, the surface is roughened by means of sandblasting. The standard test programme concerned was first proposed by Huschek (2007). Table 7.3 summarises the results of the tests involved. Except for the very smooth surface, and for the one involving texturing by means of burlap, the friction coefficient of the SHCC surface appears to be sufficient for roads that are subjected to heavy traffic. This conclusion is based on the comparison with values obtained for other concrete surfaces that are used in road construction.

Table 7.3 Results of the friction tests on SHCC surfaces performed at TU Dresden (Wellner et al. 2016)

Surface texturing	Original friction coefficient	Friction coefficient after	
		90,000 polishing cycles	180,000 polishing cycles
Steel broom	0.499	0.469	0.406
Burlap	0.247	0.328	0.337
Smooth	0.633	0.560	0.466
Very smooth	0.398	0.387	0.343
Grinding	0.561	0.580	0.438

7.4 Recommendations for Durability Design and Conclusions

For the durability design of SHCC subjected to abrasion, the following recommendations are made:

1. The material's tensile strength should be comparatively high while sufficient ductility for the respective service conditions must be ensured. SHCC with a tensile strength below 3 MPa should not be used for structures with surfaces that are subjected to abrasion.
2. Before application, the abrasion resistance of the SHCC should be determined experimentally. Recommended test methods are the method proposed by *Wehner/Schulze* and the Michigan Test Method 111 (MDOT 2001). Both these methods include the measurement of the friction coefficient as influenced by simulated road traffic.

SHCC has proven capable of meeting the technical requirements for surfaces exposed to road traffic. However, because of the lack of long-term experience and the rareness of experimental results, the applicability of the chosen SHCC composition should be tested in each individual case.

In the specific case of hydro-abrasion, there is not yet enough information to be able to draw reliable conclusions on SHCC durability under such conditions of exposure.

References

DIN EN 13813:2003-01 (2003). Screed material and floor screeds – Screed materials – Properties and requirements. Deutsches Institut für Normung, January 2003.

DIN 52108:2010-05 (2010). Testing of inorganic non-metallic materials - Wear test using the grinding wheel according to Böhme - Grinding wheel method. Deutsches Institut für Normung, May 2010.

Huschek, S. (2007). Straßengriffigkeit im Rahmen der Deutsch-Französischen Zusammenarbeit auf dem Gebiet des Straßenwesens (Road grip in the framework of German–French cooperation in the field of road engineering). Bericht (Research Report) zum Forschungs- und Entwicklungsvorhaben 04.192/2003/CRB des Bundesministeriums für Verkehr, Bau und Stadtentwicklung, Abteilung Straßenbau, Straßenverkehr, Bonn, Wirtschaftsverlag NW, Bremerhaven, Germany.

Li, V.C., Lepech M. (2004). Crack resistant concrete material for transportation construction. 83rd Annual Meeting of the Transportation Research Board, Washington D.C., USA, CD ROM, paper 04-4680.

Li, V.C., Li, M. (2008). Durability performance of ductile concrete structures. T. Tanabe, K. Sakata, H. Mihashi, R. Sato, K. Maekawa, H. Nakamura (eds.), Creep, Shrinkage and Durability Mechanics of Concrete and Concrete Structures, Proceedings of the 8th International Conference on Creep, Shrinkage and Durability of Concrete and Concrete Structures (CONCREEP 8), September 30 - October 2, 2008, Ise-Shima, Japan, Taylor & Francis Group, London, 2009, Volume 1, pp. 761-767.

Michigan Department of Transportation (MDOT) (2001). Michigan Test Method 111 – Determining an Aggregate Wear Index (AWI) by Wear Track Polishing Tests. Lansing, Michigan.

Rohde, D. (2012). Experimentelle Untersuchungen zur Dauerhaftigkeit und zum Verschleißverhalten hochduktiler Betone. Diploma thesis, Institute of Construction Materials, Technische Universität Dresden, Germany.

Van Zijl, G.P.A.G., Wittmann, F.H. eds. (2011). Durability of strain-hardening fibre-reinforced cement-based composites (SHCC). State-of-the-Art Report prepared by Subcommittee 2 of RILEM Technical Committee 208-HFC, Springer.

Wagner, C. (2007). Nachverfestigendes zementgebundenes Material für die Sanierung gerissener Betonoberflächen (Strain-hardening cement-based material for the repair of cracked concrete surfaces). Master's thesis, Leipzig University of Applied Sciences, Germany.

Wellner, F., Mechtcherine, V., Leischner, S., Butler, M., Paschke, K., Müller, S., Reinhard, U., Blasl, A., Kraft, J. (2016). Neuartige Baustoffe und Bauweisen für schwerste zukünftige Belastungen aus Klima und Verkehr: „Faserbeton" (New structural materials and construction methods for heaviest future loads resulting from climate and traffic: "Fiber concrete"). Research Report, Projekt FE 09.0173_2011_HRB for the Federal Highway Research Institute (BASt), Technische Universität Dresden, Germany.

Chapter 8
Behaviour of Bonded SHCC Overlay Systems

Volker Slowik, Mladena Luković, Christian Wagner and Gideon P.A.G. van Zijl

Abstract Strain-hardening cement-based composite (SHCC) has proved to be a suitable material for repair layers on concrete substrate. Bonded SHCC overlays may bridge cracks in the substrate, and normally exhibit an enhanced resistance to drying shrinkage. In addition to the SHCC material properties, the interface behaviour has a significant influence on the failure process. By varying the interface roughness and bond strength, it is possible to attain a balance between the debonding of the interface and SHCC cracking, and to ensure both monolithic mechanical behaviour of the structure and sufficiently small crack widths in the SHCC overlay. The smaller these crack widths are, the lower the permeability and the greater the self-healing potential of the cracks. In the case of mechanical loading, a rather weak bond may be beneficial, as the crack widths in the SHCC tend to be smaller under such conditions. In the case of drying shrinkage of the SHCC overlay, however, weak bonding may lead to large debonded interface areas, making the use of SHCC inadvisable in such instances.

Keywords Crack bridging · Shrinkage · Bond · Mixed mode · Interface modelling

8.1 Introduction

Repair layers on cracked concrete surfaces are probably one of the most important applications of strain-hardening cement-based composites (SHCC). The suitability of SHCC in this respect can mainly be attributed to the crack-bridging capability of

V. Slowik (✉) · C. Wagner
Leipzig University of Applied Sciences, Leipzig, Germany
e-mail: volker.slowik@htwk-leipzig.de

M. Luković
Delft University of Technology, Delft, The Netherlands

G.P.A.G. van Zijl
Stellenbosch University, Stellenbosch, South Africa

G.P.A.G. van Zijl and V. Slowik (eds.), *A Framework for Durability Design with Strain-Hardening Cement-Based Composites (SHCC)*, RILEM State-of-the-Art Reports 22,
DOI 10.1007/978-94-024-1013-6_8

SHCC repair layers and to their chemical compatibility with the cement-based substrate. Due to strain hardening, SHCC overlays are capable of bridging cracks with comparatively large opening displacements in the substrate. For characterising and modelling the phenomenon, it is not sufficient to consider only the material behaviour under uniaxial tension. The mechanical behaviour of the interface between the SHCC repair layer and the substrate also requires consideration, as it has a significant influence on crack widths and spacing in the SHCC. Consequently, the bond properties will affect the durability of the overlay.

Bonded SHCC overlays may be applied to repair existing structures, but they may also form part of new construction. In both cases, the durability of the bonded connection is essential for the structural durability of the whole. Refer for instance to Chap. 1, Sect. 1.5.1, related to patch repair, as well as Sects. 1.5.3 and 1.5.4, where connecting elements in highway bridge systems rely on the interfacial bond between SHCC and reinforced concrete to transfer forces while maintaining structural continuity for control of the superstructure and for substructure durability. In all these cases, the performance has been reported to be sound, even after several years of traffic and environmental exposure, including de-icing salt practice.

The high ductility of the SHCC results in a specific interface behaviour between this material and ordinary concrete. For this reason, the bond models proposed in the literature for concrete/concrete interfaces cannot be applied directly. Experimental investigation of the behaviour of bonded SHCC overlays was required, as was the proposal of appropriate interface models to account for the characteristic material properties of the SHCC.

The mechanical behaviour of a bonded SHCC overlay on a concrete surface with an opening crack is characterised by a combination of delamination and overlay cracking. Lim and Li (1997) were the first to report on the so-called kink-crack trapping. During continuous crack opening in the substrate, delamination starts from the edges of the respective crack. Subsequently, the locally delaminated overlay will also crack because of the relatively low SHCC tensile strength. These cracks originate from the ends of the de-laminated range and propagate into the SHCC. In this way, the damage is drawn into the SHCC overlay, where the cracks are 'trapped', temporarily stopping the delamination. Due to the strain hardening, the resistance to further crack propagation in the SHCC eventually becomes higher than the resistance to further delamination. Consequently, the delamination proceeds until the next 'trapping' occurs. The process leads to multiple cracking and to the activation of the SHCC ductility. However, such behaviour requires an appropriate balance between bond and overlay properties. If the bond strength is too high, the crack-bridging capability of the SHCC overlay may not be fully activated. Therefore, in contrast to the common approaches taken to repairing concrete, the presence of high bond strength in SHCC overlays is not always beneficial.

Multiple cracking in SHCC overlays leads to comparatively small crack width and, consequently, to reduced permeability (see Chap. 2). This is beneficial for the durability of both substrate and overlay. It has to be taken into account, however, that even small cracks in SHCC are almost immediately filled with water when capillary absorption takes place (Wang et al. 2014). By using water-repellent agents, the resistance to fast capillary absorption may be increased (see Sect. 8.6). Other benefits of the small crack widths in SHCC overlays are the greater self-healing potential and, presumably, improved abrasion resistance of the material.

8.2 The SHCC-Substrate Interface

8.2.1 Substrate Crack Bridging

Kamada and Li (2000) concluded, on the basis of their experimental results, that a smooth interface between SHCC overlay and concrete substrate should be preferred to the presence of rough interfaces. They conducted bending tests of concrete beams, with a continuous gap imitating the presence of a crack in the substrate (thus undertaking a reflective cracking test). During the test, the overlay was located in the compression zone. Lim and Li (1997) had previously used a similar experimental setup. Kunieda et al. (2004) also performed bending tests of SHCC-concrete beams with an artificial crack in the concrete substrate. However, in their case, the SHCC overlay was in the tension zone of the beams. The tests mentioned, as well as zero-span elongation tests conducted by Wagner et al. (2008) and Wagner and Slowik (2011), could demonstrate the crack-bridging capability of SHCC overlays. The latter source contains, in addition, the results of bending tests that were performed on pre-cracked reinforced concrete beams, with an SHCC repair layer applied to the tension face. The experiments, and numerical simulations of the same (Wagner et al. 2008), allowed for the study of the crack-bridging effect, due to multiple cracking in the repair layer (see Fig. 8.1). For simulating the interface behaviour, a Coulomb-friction-type failure criterion implemented in ATENA by Červenka Consulting, Prague was used. In the experiments, as well as in the simulations, the crack widths on the SHCC surface were found to be significantly smaller than were those in the concrete substrate. On the basis of their experimental results, Wagner and Slowik (2011) concluded that crack-opening displacements of up to 0.6 mm in a substrate can be bridged by a 3 cm thick SHCC repair layer. Digital image processing yielded crack widths below 50 μm on the SHCC surface. However, when the crack-opening displacement in the substrate exceeded a limit of approximately 0.6 mm, the SHCC repair layer exhibited localised cracking with rapidly increasing maximum crack width on the surface.

8.2.2 Mesoscale Modelling Towards Interface Optimisation

Luković et al. (2014b) used a mesoscale approach for investigating the fracture behaviour of bonded SHCC overlays. In their lattice models, the distribution of

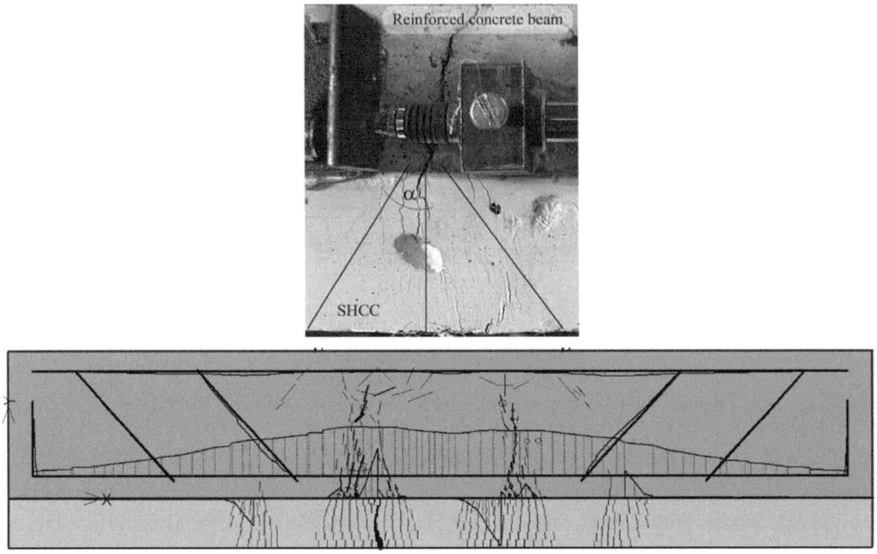

Fig. 8.1 Multiple cracking in an SHCC repair layer on a pre-cracked reinforced concrete beam subjected to bending, side view; *top* crack pattern in the SHCC repair layer (Wagner and Slowik 2011); *bottom* results of a non-linear finite element analysis, with ATENA using interface and smeared crack models (Wagner et al. 2008), tension zone at the bottom, repair layer underneath the x-axis, debonding cracks in the interface not shown, graph above the longitudinal reinforcement showing steel stress

discrete fibres, and the actual interface geometry, with its roughness, may be directly reproduced. Such reproduction allows for the making of numerical experiments in which individual influences on the fracture behaviour can separately, and independently, be investigated. Although care must be taken when quantitatively evaluating the simulation results based on certain model limitations, an improved understanding of the several effects of fracture behaviour can be gained with the help of the numerical experiments concerned. First, the SHCC repair material was tailored by means of fitting the numerical results of simulated singe fibre pull-out tests and uniaxial tension tests to experimental observations. Using the optimised 'numerical' SHCC, three-point bending tests of repair-substrate systems were simulated in 3D. The substrate specimens, which had the dimensions of $10 \times 15 \times 60$ mm^3 (with a height of 10 mm), were initially uncracked. A SHCC repair layer of 5 mm thickness was applied to the tension face. Note that, due to computation costs, specimen sizes were limited. For comparison, a repair material without fibres has also been tested. Since the surface roughness was expected to strongly influence the bond behaviour between substrate and overlay, it was varied. Figure 8.2 shows three surface profiles that were used in the numerical experiments reported on by Luković et al. (2014b). A more detailed investigation into the impact of differently grooved substrate surfaces was published by Luković et al. (2013).

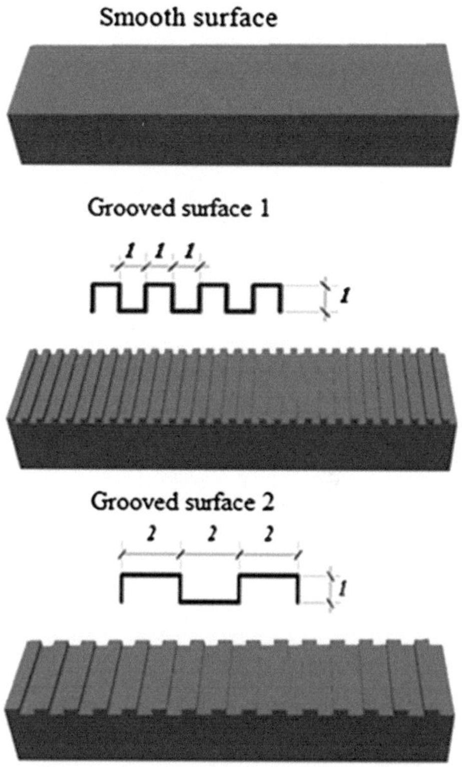

Fig. 8.2 Surface profiles considered in the numerical experiments by Luković et al. (2014b)

In Fig. 8.3, the force-deflection diagrams are shown for the three surface profiles, with and without fibre reinforcement of the overlay. The brittle post-peak behaviour of the non-reinforced material resulted from the formation of a single crack, whereas multiple cracking occurred in the SHCC overlays. The surface profile appears to have had no significant influence on the global flexural response obtained. The crack patterns, however, differ for the three surface profiles (see Fig. 8.4). A smooth surface results in the formation of a relatively long debonding length. In the case of SHCC, the presence of such length leads to the lengthening of the zone of distributed cracking and, consequently, to the presence of additional cracks that are beneficial to the structural durability involved (since the presence of additional cracks means that the cracks are relatively fine). A similar observation was made in experiments by Kunieda et al. (2004). In Fig. 8.4, it may also be seen that, in the case of the presence of a smooth surface, the cracks in the SHCC do not propagate directly into the substrate, so that cracking of the latter is relatively localised. Such localisation of cracks is a consequence of the missing interlocking at the interface. Due to the relatively long debonded length, the deflection at final failure also tends to be relatively large in the case of the presence of a smooth surface.

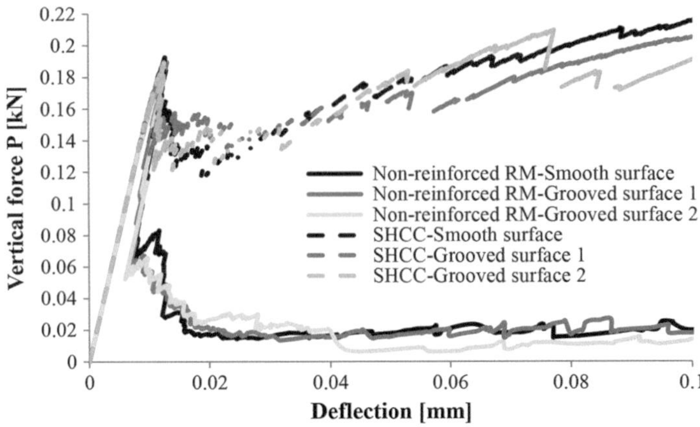

Fig. 8.3 Load-deflection curves of the simulated three-point bending tests; lattice model simulations (Luković et al. 2014b)

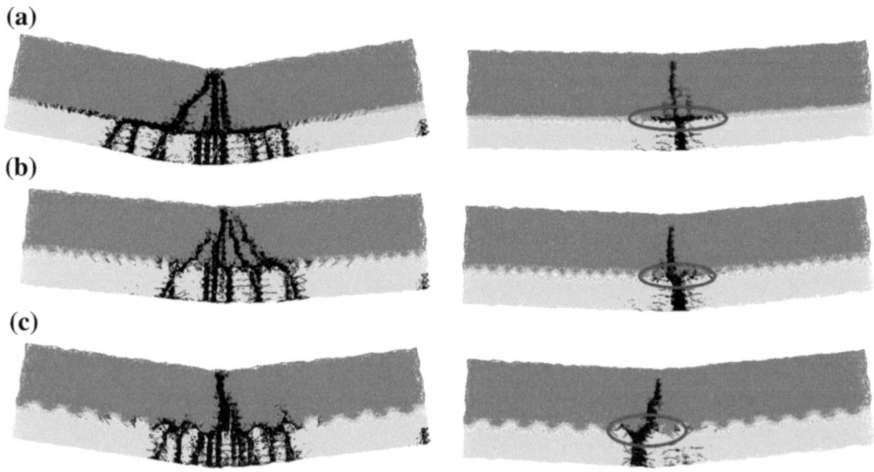

Fig. 8.4 Crack patterns obtained in the simulated three-point bending tests, with SHCC repair layer at the bottom; lattice model simulations; *left* final crack pattern; *right* crack pattern in the case of a repair layer without fibre reinforcement, at 0.02 mm deflection; **a** smooth surface, **b** grooved surface 1, **c** grooved surface 2 (Luković et al. 2014b)

As far as the influence of the bond strength is concerned, Luković et al. (2014b) obtained similar results as for the varying surface roughness. The lower the bond strength, the longer the debonded length, and the more distributed the cracking in the SHCC will be, while the substrate cracking tends to be relatively localised. By means of varying the surface roughness or bond strength, substantially changing of the failure behaviour appears to be possible. The presence of smooth interfaces or low bond strength results in extensive debonding, with multiple cracking in the repair layer, but with localised cracking in the substrate. In contrast, rough interfaces or high bond strength result in fewer cracks in the repair layer, and the presence of distributed cracking in the substrate. It has to be highlighted, however, that there is a threshold (or minimum) bond strength that might cause the final failure mode to change and result in complete delamination in the overlay system (Luković 2016).

For comparison with the above simulations, Luković et al. (2014b) performed three-point bending tests of composite beams with the dimensions $40 \times 40 \times 160$ mm^3. On top of a mortar substrate, a 15 mm thick SHCC repair layer was applied. The size of the grooves in the experiments (being 3 mm deep, and 3 mm wide) had the same ratio to the specimen size as in the numerical simulations. The intention was to compare the global fracture behaviour and crack processes in the experiments to those in the numerical simulations. There was no intention to compare the absolute values of the results obtained. The crack patterns were evaluated by means of digital image correlation (DIC). When compared to the simulation results, the same sequence of crack propagation, and the same effects of the interface properties on both the global force-deflection curves and the crack patterns, were found.

The behaviour of SHCC overlays on a concrete substrate with a single crack was also simulated with lattice models (Luković et al. 2014b). The intention was to find out how a crack in the substrate reflects into the applied overlay (in the form of reflective cracking). Simulated beams with the dimensions $12 \times 12 \times 60$ mm^3 and a 6 mm overlay were subjected to three-point bending. The beam, namely the substrate, had a gap at midspan, and bending was applied, such that the gap was at the tension face, as had been the case in experiments carried out by Kamada and Li (2000). (Note that the loading direction in the bending tests with pre-cracked substrate performed by Kunieda et al. (2004) and by Wagner and Slowik (2011) was different). Figure 8.5 shows the obtained crack patterns. In case of an unreinforced repair mortar, a single crack, extending from the pre-existing crack in the substrate, is formed in the overlay (see Fig. 8.5, top). When SHCC is applied, however, the crack-bridging effect may be demonstrated (see Fig. 8.5, bottom). A smooth surface results in a relatively long debonded length, and in a relatively

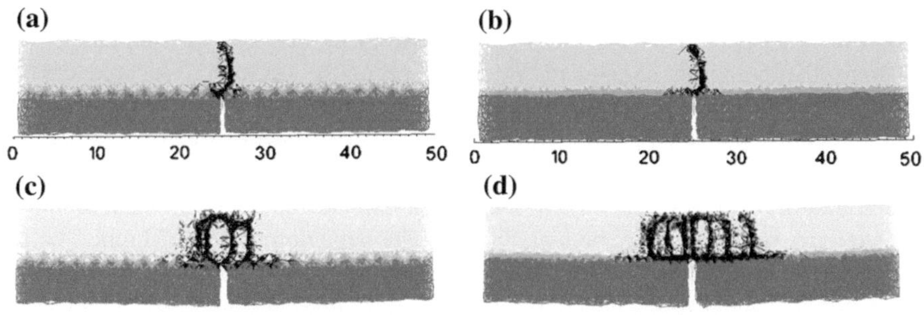

Fig. 8.5 Crack patterns obtained in the simulated three-point bending tests with an artificial crack in the substrate; lattice model simulations; *top row* crack pattern in case of a repair layer without fibre reinforcement; *bottom row* crack pattern in case of an SHCC repair layer; **a**, **c** grooved surface, **b**, **d** smooth surface (Luković et al. 2014b)

distributed crack pattern in the SHCC overlay, in comparison to the features of the grooved surface. In both cases, the fracture sequence is characterised by debonding, overlay cracking, crack arresting in the strain-hardening overlay, namely ‘trapping’, and continued debonding, until the next overlay crack is initiated.

8.3 Drying Shrinkage and Interface Behaviour

In the previously described numerical simulations with lattice models, drying shrinkage was not taken into account. A smooth interface or low bond strength was found to be of potential benefit to the structural behaviour, due to the presence of a relatively distributed crack pattern in the SHCC overlay. However, when differential shrinkage is considered, meaning that when there is a higher shrinkage strain in the overlay than in the substrate, care should be taken in recommending the use of certain interface properties. In the case of a smooth or low-strength interface, drying may lead to large and uncontrolled debonding and, possibly, to the failure of the complete overlay already before the strain-hardening behaviour can be utilized (Luković et al. 2014a, b). Therefore, sufficient bonding should be ensured in order to mitigate the damage that might occur due to drying shrinkage. That SHCC exhibits comparatively high drying shrinkage strains must be considered. The presence of such high strains may be attributed to the high binder content, and to the lack of large aggregate particles, in the material concerned. Nevertheless, Luković et al. (2014a) could demonstrate in lattice model simulations that drying shrinkage cracks in SHCC overlays have smaller widths than have those that are formed in an unreinforced reference mortar with the same bond strength and substrate surface roughness, when they are subjected to the same environmental conditions. Due to the strain hardening of the SHCC, the number of cracks is likely to be larger, and the crack spacing smaller, in the latter material. An increasing interface roughness or bond strength, or else a reduction of the overlay thickness from 40 to 20 mm was also found to lead to a more distributed crack pattern in the simulations (Luković et al. 2014a). Figure 8.6 shows the crack patterns obtained in lattice models that were subjected to simulated drying at the upper surface.

Strain hardening is concluded as being generally beneficial for the drying shrinkage resistance of cement-based overlays, since distributed fine cracking occurs, rather than does the formation of a few relatively wide cracks. As far as the interface properties are concerned, contradictory trends exist for the mechanical and hygral loading of SHCC overlay systems (Van Zijl and Stander 2009). Smooth, or low-strength, interfaces may be beneficial for the behaviour that is exhibited under conditions of mechanical loading. Under the action of drying shrinkage, however, rough interfaces and relatively high bond strength appear to be more suitable features. The same conclusion is drawn for the type of damage that is induced by means of the continued corrosion of steel reinforcement that is embedded in SHCC overlay systems, in which high substrate roughness and good bond are also beneficial properties (Luković et al. 2014c).

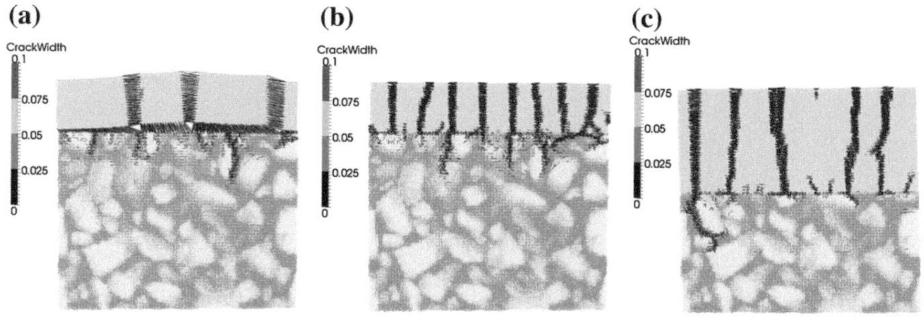

Fig. 8.6 Crack patterns obtained under simulated drying conditions at the upper surface of an SHCC overlay; lattice model simulations; smooth interface; **a** overlay thickness of 20 mm, bond strength of 1 MPa, **b** overlay thickness of 20 mm, bond strength of 3 MPa, **c** overlay thickness of 40 mm, bond strength of 3 MPa (Luković et al. 2014a)

8.4 Interface Mode I and II Behaviour

Whereas the previously mentioned tests and numerical simulations were conducted for the purposes of investigating the mechanical behaviour of a complete SHCC overlay system, experiments for separately investigating the mode I (tension) and mode II (shear) behaviour, respectively, have also been reported. The objective of the experiments in question was, primarily, the identification of parameters for macroscopic interface models that may, eventually, be used in the design of overlay systems. Van Zijl and Stander (2009) conducted uniaxial tension tests, as well as shear tests, at interfaces between SHCC and ordinary concrete. The interface tensile strength, obtained for a cross-sectional area of 100×130 mm^2, was found to be significantly lower than was the uniaxial SHCC tensile strength (with the former being less than 30% of the uniaxial SHCC tensile strength, and with the ratio concerned varying, depending on the substrate surface roughness). Delamination was observed in the case of a smooth substrate, whereas, in the case of a sand-blasted substrate, failure occurred predominantly in the SHCC. Even in the last-mentioned case, the maximum stress was lower than was the SHCC tensile strength, which could be attributed to the fibre orientation running parallel to the interface, and to the local weakening due to substrate-overlay interaction. Expectedly, the SHCC failure near the interface was less brittle than was the delamination. For their shear tests, Van Zijl and Stander (2009) used push-off specimens (see Fig. 8.7), which were subjected to a compressive force that acted parallel to the interface. The size of the shear area was 100×100 mm^2, and, in numerical pre-investigations, non-rotational boundaries on both end faces were found to be suitable for the tests undertaken. No lateral confining pressure was applied. Expectedly, the interface shear resistance appeared to increase with the degree of interface roughness. The shear strength of a smooth interface was approximately one third of that which was obtained for a scraped substrate, which, in turn, was about 23% lower than was the one for a sandblasted substrate.

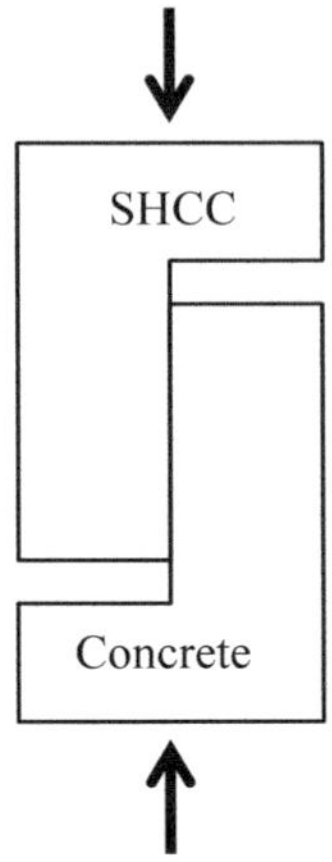

Fig. 8.7 Shear test specimen (reproduced from Van Zijl and Stander 2009)

Furthermore, a ten minutes moistening of the substrate led to a higher shear resistance than did 24 h of submergence in water. The observation was attributed to a relatively high degree of capillary suction from the fresh SHCC, leading to a relatively great interlock along the interface. The interface tensile strength appeared to be insensitive to the moistening method, whereas the mode I fracture energy was higher in the case of ten minutes of moistening. However, ten minutes of moistening tended to give rise to increased uncertainty, as the variability of the results was higher than in the case of 24 h of submergence in water.

Fracture tests at concrete-SHCC interfaces were also performed by Wagner et al. (2013, 2014). As in the experiments reported by Van Zijl and Stander (2009), the SHCC had a polyvinyl alcohol (PVA) fibre content of 2% by volume, with the uniaxial tensile strengths being approximately the same. In addition to the wedge splitting tests that were aimed at investigating the mode I fracture behaviour, Wagner et al. (2013, 2014) conducted mixed-mode fracture tests. Figure 8.8a shows the experimental set-up that was used for the wedge splitting tests (mode I). The ligament width amounted to 100 mm, and the ligament length ranged from 60 to 90 mm, with the latter not exhibiting a systematic influence on the obtained fracture properties. For the mixed-mode fracture tests, the set-up of the slant shear test was chosen (see Fig. 8.8b), which allowed to vary the tension–shear ratio by means of changing the inclination of the joint. It was, however, impossible to directly separate the damage levels that were locally reached under tension and shear, respectively. Therefore, inverse analyses of the experiments were required. In these analyses, the non-uniform stress distribution along the interface was taken into account. In contrast to the specimen geometry that was originally proposed for the slant shear test, two notches were cut into the specimens. The remaining interface area was 70 × 70 mm^2 for all inclinations α of the interface. Two reasons existed for the notches. On the one hand, surface effects on the material behaviour were to be excluded, since the local material properties in the vicinity of the formwork may differ from those in the bulk of the material. On the other hand, in

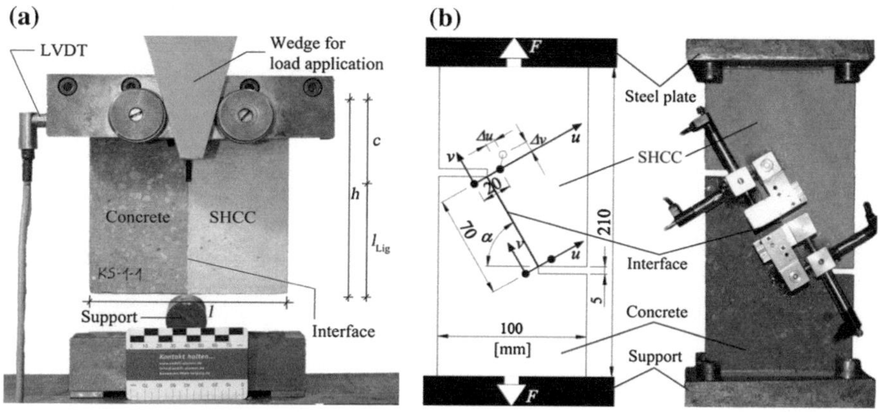

Fig. 8.8 Experimental set-ups for **a** the wedge splitting tests, and **b** the slant tension-shear tests, conducted by Wagner et al. (2013)

the absence of notches, acute-angled edges would have been formed in both halves of the respective specimen. In the regions concerned, failure of the material itself might have occurred, instead of the intended bond failure. Slant shear specimens with six different joint inclinations α (0°, 20°, 40°, 60°, 75°, and 90°) were tested. An angle of zero results in pure tension, and an angle of 90° predominantly in shear. Figure 8.8b shows a specimen with a joint inclination of 60°. At each of the notch tips, two orthogonal displacement components were measured by means of linear variable differential transformers (LVDTs).

For determination of the mode I fracture properties, inverse analyses of the wedge splitting tests were conducted, allowing for the following conclusions to be drawn (Wagner et al. 2013). If the surface roughness of the interface is low, the bond strength depends strongly on the roughness value involved. The influence of the surface roughness on the bond strength vanishes, however, when the bond strength becomes high. In such a case, cracks are formed in the SHCC, and the fibres are activated. The higher the area portion with an intact bond between SHCC and concrete, meaning that the higher the area portion with SHCC adherence, the more ductile the bond behaviour will be. Hence, the mode I fracture energy strongly depends on the portion of the fracture surface with SHCC adherence (see Fig. 8.9a). The results of the wedge splitting tests also show that, in the case of low bond strength when compared to the matrix strength, the failure is restricted to the actual interface, and the SHCC properties do not influence the bond behaviour. Wagner et al. (2013) quantified the surface roughness by means of the so-called patch test and determined the area portion with SHCC adherence by means of photogrammetric measurement. Their findings accord well with observations reported by Silfwerbrand et al. (2011), who found that a certain threshold value exists for the roughness of interfaces to bonded cement-based overlays. If the threshold value concerned is exceeded, a further increase of the roughness will no longer influence the bond

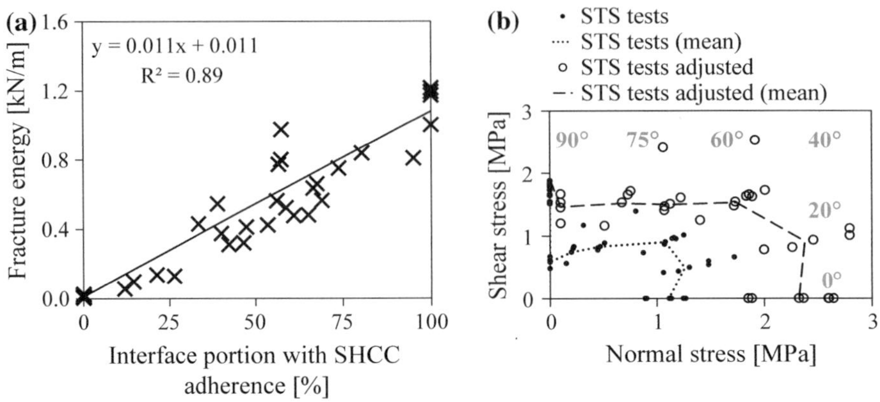

Fig. 8.9 **a** Fracture energy determined by means of inverse analysis of wedge splitting tests, and **b** ultimate stresses in the slant tension-shear (STS) tests (Wagner et al. 2013)

strength involved. According to Silfwerbrand et al. (2011), the threshold roughness is in the order of magnitude of that of typical sandblasted concrete surfaces.

Figure 8.9b displays results of the slant shear tests, meaning of the tests under combined tension and shear (mixed-mode) loading (Wagner et al. 2014). For different interface inclinations, the ultimate stresses are shown. The dots (that are labelled as STS tests) correspond to the normal and shear stresses that are obtained under the assumption of a uniform stress distribution in the interface. The circles (labelled as STS tests adjusted) represent the results of an improved evaluation, in which the stress distributions in the interfaces were determined by means of linear-elastic finite element analyses. Although an exact evaluation of the test results requires inverse analyses of the experiments, the curves in Fig. 8.9b show the tendency of an elliptic failure surface. The latter is considered in the interface model described below.

8.5 Modelling of the Interface Behaviour

Van Zijl and Stander (2009), as well as Wagner et al. (2014), proposed computational models that would allow for reproduction of the results of mode I, mode II, or mixed-mode tests, respectively, and for the identification of material parameters that could not be directly retrieved from the tests. The efforts involved were aimed at a suitable model that may eventually be utilised for the design of SHCC overlay systems. Van Zijl and Stander (2009) used a failure surface comprising a Coulomb-friction model that was limited by means of a tension cut-off and a compressive cap (see Fig. 8.10). Softening in tension and shear is based on the respective fracture energy involved. When simulating the performed tests, the compressive cap was at no stage reached. While the tensile strength, the cohesion,

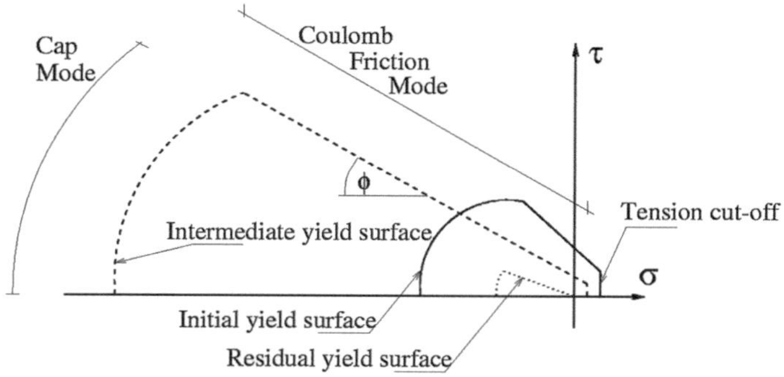

Fig. 8.10 Interface model used by Van Zijl and Stander (2009)

and the two fracture energies were directly derived from the test results, the dilatancy model parameters were determined by means of inverse analysis, meaning by fitting the simulation results to the results from the shear tests, including the measured normal uplift. Cracking of the concrete substrate was considered by using a total strain rotating crack model. For the SHCC, a uniaxial stress–strain curve, incorporating both hardening and softening, has been adopted. With the given model assumptions, the average measured response in the tension and shear tests could be reproduced (Van Zijl and Stander 2009). However, the assumed exponential tension-softening curve appeared to be imperfect for describing the mode I post-peak behaviour of the interface concerned.

After having identified the material parameters by means of analysing the tension and shear tests, the model was used for simulating flexural tests with SHCC overlays (see Fig. 8.11). In the experiments, the concrete substrate had an artificial crack, and a 50 mm delamination length was created by applying insulating tape prior to the SHCC overlay casting (Van Zijl and Stander 2009). The global flexural response obtained in the simulations was very close to that which was measured in the experiments. As far as the fracture behaviour was concerned, the effects of debonding and large strain, due to the multiple cracking of the SHCC overlay, could be reproduced in the simulations (see Fig. 8.11).

Wagner (2016) and Wagner et al. (2014) followed a similar strategy for modelling the bond behaviour of SHCC overlays on the structural (i.e. macroscopic) level, but implemented some refinements with respect to the model used by Van Zijl and Stander (2009). The refinements concerned were mainly related to the failure surface, and to the softening curves. So as to account for the specific material behaviour of SHCC, the following requirements were formulated for the interface model:

i. Simulation of combined normal and shear loading has to be possible, in terms of which the softening behaviour in the two different loading directions may differ.

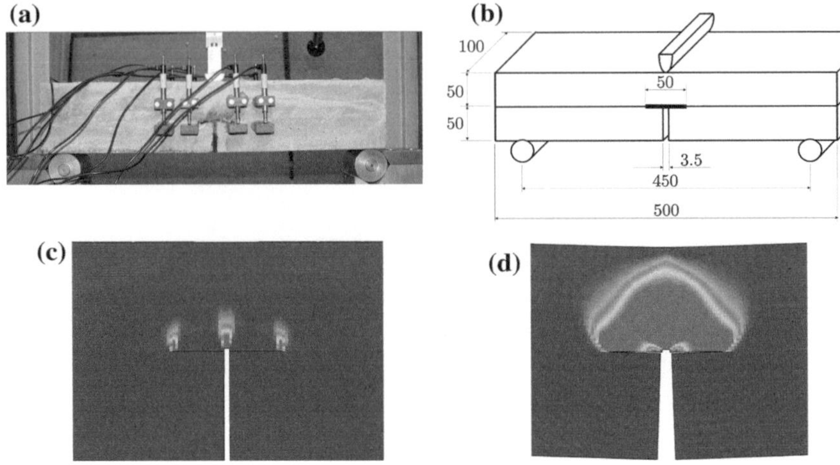

Fig. 8.11 **a**, **b** Flexural test of an SHCC overlay on concrete substrate and finite element simulation results of the computed equivalent strain, at **c** 50% of the peak load, and **d** peak load (Van Zijl and Stander 2009)

ii. Both softening curves may be multilinear, so as to account for the specific SHCC material behaviour.
iii. The residual strength under cyclic loading depends on the damage level involved.
iv. An interaction of the two loading directions is possible, in terms of which reversible, as well as irreversible, phenomena may occur.
v. The shear resistance depends on the shear strength, and on the compressive normal stress in the interface, as a consequence of the Coulomb friction.
vi. The roughness of the interface results in dilatancy.
vii. Tangential displacements reduce the macroscopic roughness and, consequently, the dilatancy angle.
viii. A reduction of the dilatancy angle may also occur as a result of compressive stresses.
ix. Damage under tension or shear (i.e. softening) is independent of the damage caused by a reduction in the dilatancy angle.

The description of the interface model requires the consideration of two different scales. Tensile strength and cohesive shear strength are macroscopic properties, with the coupling of the two loading directions also being described on the macroscale. On the mesoscale, a geometrical surface model is used for modelling the dilatancy effect and the change in the dilatancy angle due to damage. From the mesoscale surface model, the maximum compressive stress in the interface and the friction stresses can also be derived. Figure 8.12a shows the interface. It is assumed that the surface model consists of a series of identical saw teeth with the inclination ψ. The inclined slopes of the mesoscale surface model have the friction angle ϕ_R.

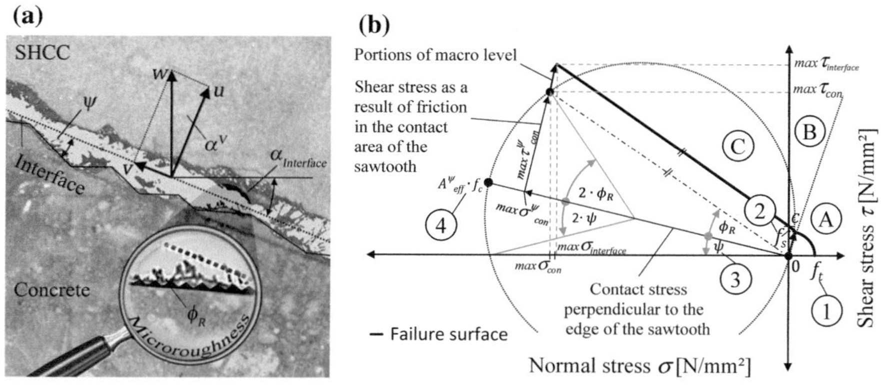

Fig. 8.12 **a** Modelling of an idealised interface, and **b** failure surface (Wagner et al. 2014)

Figure 8.12b shows the failure surface with the following three regions: A (tension-shear), B (transition), and C (compression-shear). The Coulomb line in the compression-shear region (C in Fig. 8.12b) is combined with an ellipse in the tension-shear region (A). The elliptical shape corresponds to the experimental observations. The transition zone (B) between the two aforementioned functions depends on the dilatancy angle, as well as on the tension and shear strength of the interface concerned. Equation 8.1 defines the failure surface with its three regions:

$$F(\sigma,\tau)=\begin{cases}\left(\frac{\sigma}{f_t}\right)^2+\left(\frac{\tau}{f_s}\right)^2-1=0 & \text{for } 0\le\alpha^\sigma\le\alpha^\sigma_{con} \quad (A)\\ \tau-\sqrt{f_s^2-\frac{f_s^2\cdot\sigma^2}{f_t^2}}+\frac{\tan\phi_{eff}}{(\sigma-\sigma_{Limit})^{-1}}=0 & \text{for } \alpha^\sigma_{con}<\alpha^\sigma<\frac{\pi}{2} \quad (B)\\ \tau+\tan\phi_{eff}\cdot\sigma-c=0 & \text{for } \alpha^\sigma\ge\frac{\pi}{2} \quad (C)\end{cases} \tag{8.1}$$

with:

σ normal stress in the interface
τ shear stress in the interface
f_t tensile strength
f_s shear strength
c cohesion
ϕ_{eff} effective friction angle (being the sum of the friction angle ϕ_R and of the effective dilatancy angle ψ_{eff})
σ_{Limit} the limit tensile stress at which the contact stresses at the saw teeth's surfaces are activated
α^σ the direction of the stress resultant
α^σ_{con} the contact angle in the stress space

The contact angle depends on the state of deformation, as well as on the effective stiffnesses K_u and K_v in the two loading directions:

$$\alpha_{con}^{\sigma} = \tan^{-1}\left(\frac{K_v}{K_u} \cdot \frac{v}{u_{Dil}(v)}\right) \tag{8.2}$$

with:

v tangential displacement
$u_{Dil}(v)$ displacement normal to the surface at angle ψ_{eff}

At the intersection of regions A and B (see Fig. 8.12b), the resistance of the interface changes from the tension-shear strength to a tension-shear-contact strength. The point in question separates the damage that results from softening under shear and tension from the damage that results from the reduction of the surface angle. In contrast to other failure surfaces (Coulomb, or Carol et al. 1997), a changing friction angle will, in the present instance, not affect the strength in the tension-shear region.

The failure surface will change as a result of the following four mechanisms, which are accordingly numerated in Fig. 8.12b:

(1) Damage by reduction of the tensile strength (tension softening)
(2) Damage by reduction of the shear strength (shear softening)
(3) Damage by reduction of the surface angle, i.e., dilatancy angle ψ (surface damage)
(4) Limitation of the compressive strength, and local crushing at the saw teeth

The geometrical surface model with the saw teeth is shown in Fig. 8.13a. A model of this type has already been used for SHCC by Zhang et al. (2011). The reduction of the tensile and shear strengths is described by means of multilinear softening curves (see Fig. 8.13b). This part of the figure also shows the coupling of the two loading directions by $w_{eff} = \sqrt{u^2 + v^2}$, based on the assumption that, in the case of fibre bridging, no further damage occurs as long as w_{eff} is not increased. The above means that the free end of a partially pulled-out and stressed fibre may describe a circle under a changing state of displacement without further damage.

For the surface damage, two different situations have to be considered. Firstly, the saw teeth may be physically damaged due to shear displacement in the interface. Secondly, a normal opening displacement may result in a reduction of the 'efficiency' of the saw teeth, due to the reduction of the contact stress. The 'efficiency' (i.e. the effect of the saw teeth) is reactivated, however, if the normal displacement is reversed. For modelling the damage that is experienced under conditions of shear displacement, a proposal developed by Carol et al. (1997) was used (see Fig. 8.13c). Thereby, the effective dilatancy angle ψ_{eff} can be calculated on the basis of a damage parameter α_{Dil}:

$$\tan\psi_{eff} = \begin{cases} \tan\psi_0 \cdot \left(1 - \frac{e^{\alpha_{Dil}} \cdot v}{(e^{\alpha_{Dil}} - 1) \cdot v + v_{crit}}\right) & \text{for } 0 \le v \le v_{crit} \\ 0 & \text{for } v_{crit} \le v \end{cases} \tag{8.3}$$

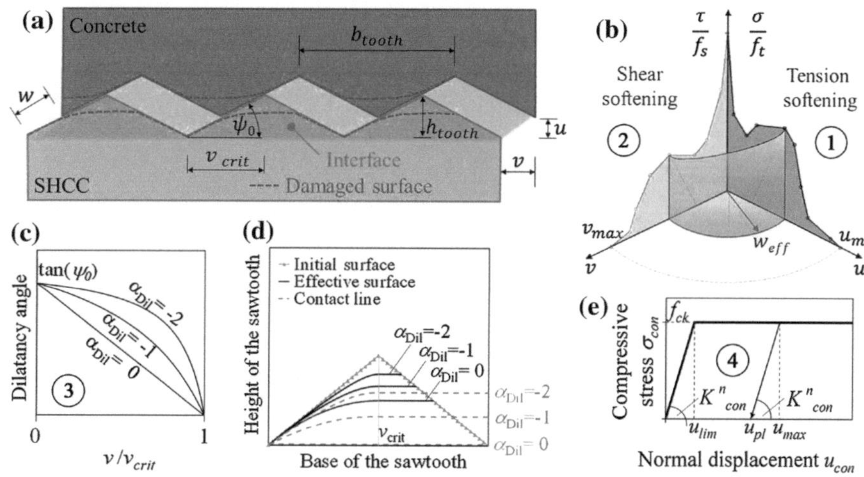

Fig. 8.13 Surface model and damage mechanisms (Wagner et al. 2014)

The integration of the above-mentioned function yields the contour of the damaged saw tooth, as is shown in Fig. 8.13d for three different damage parameters α_{Dil}. The contact lines in the part of the figure concerned indicate under which degree of displacement contact stresses are present. It is assumed that the saw teeth at the opposite sides are equally damaged. Without contact stresses, the interface is subjected to tension and shear only.

The compressive strength of the interface is the product of the material's compressive strength and the contact area portion in the interface. The resulting limitation of the interface's compressive strength is the fourth damage mechanism. For the material's compressive behaviour, the elasto-plastic model, shown in Fig. 8.13e, was adopted.

Consequently, the model proposed by Wagner (2016) and by Wagner et al. (2014) requires the following interface parameters: the softening curves for tension and shear, including the corresponding unloading slopes, the compressive strength f_{ck}, the original (i.e. undamaged) dilatancy angle ψ_0, the friction angle of the surface ϕ_R, the damage parameter α_{Dil} (see Fig. 8.13c), and the critical tangential displacement v_{crit} (see Fig. 8.13a).

Figure 8.14 shows the inverse analysis results for three slant tension-shear tests with 40° inclination of the interface obtained with the proposed model. In Fig. 8.14a, the solid lines are the experimentally obtained load-displacement curves, whereas the symbols represent the inverse analysis results for the respective experiment. An evolutionary optimisation algorithm allowed for very good fits of the numerical results to the experimental ones. The softening curves obtained for the two loading directions are presented in Fig. 8.14b. Both sets of curves have a characteristic shape, with the deviations between the individual curves being acceptable when the scatter of the experimental data is taken into consideration.

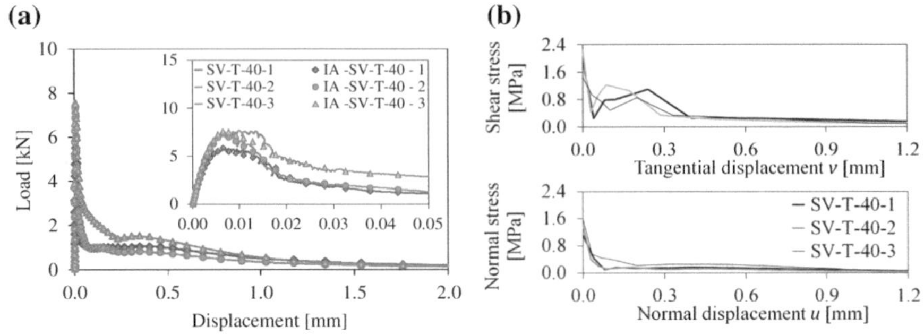

Fig. 8.14 **a** Inverse analysis results for three slant tension-shear tests, 40° inclination of the interface, experimentally and numerically determined load-displacement curves up to 2 mm, and up to 0.05 mm displacement (zoomed); **b** softening curves for tangential and normal displacement, respectively (reproduced from Wagner 2016)

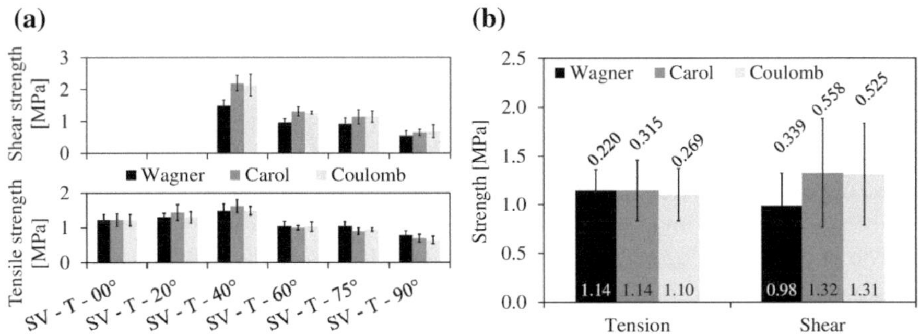

Fig. 8.15 **a** Mean strength values and standard deviations for different inclinations of the interface; **b** average strength values for all inclinations, with the corresponding standard deviations (reproduced from Wagner 2016)

For different inclinations of the interface, almost the same tensile strength values were obtained (see Fig. 8.15a, bottom). The shear strength values, however, appeared to be influenced by the interface's inclination (see Fig. 8.15a, top), although the deviations are smallest for the model proposed by Wagner (2016). For the shallow inclinations of 0° and 20°, the shear softening does not have a significant influence on the fracture process. Therefore, inverse analyses of tests with these inclinations will not yield reliable shear properties.

For comparison, inverse analyses were also carried out with other failure surfaces, in particular with the one proposed by Carol et al. (1997), and with the Coulomb criterion modified by means of a tension cut-off. When the results were averaged out over all the interface inclinations concerned, the failure surface proposed by Wagner (2016) was found to yield the smallest standard deviation (see Fig. 8.15b). Expectedly, the proposed model tends to yield lower shear strength values when it is compared to the other models. In addition to the actual cohesive shear strength, the geometrical shape of the interface assumed in the present instance also contributes to the resistance to tangential displacements (i.e. to the total shear strength).

8.6 Protective Function of SHCC Overlay Systems

Whereas the interfacial bond and its characterisation has been the focus of the current chapter, reference to the durability of the overlay system and, in particular, to the substrate, is relevant in the context of the STAR at hand. Although other chapters deal with such particular aspects as ingress (Chap. 2) and corrosion (Chap. 9), reported laboratory results of ingress through SHCC overlays into the substrate are referred to in the present section.

Schröfl et al. (2015) tested 15 mm thick bonded SHCC overlays cast onto 60 mm deep, 300 mm wide, and 1 m long R/C beams of class C25/30 concrete. The R/C beams were pre-cracked in a four-point bending arrangement with a span of 900 mm, causing cracks of a length up to 50–60% of the beam depth of 60 mm. Substrate preparation included sandblasting of the cracked concrete surface, which was subsequently sprayed with about 15 mm thick SHCC overlay. After 28 days of curing, the overlay system was again subjected to four-point bending, whereby multiple cracks arose in the SHCC overlay throughout its full 15 mm thickness. Figure 8.16a shows a schematic view of the overlay concerned.

Two specimen types were prepared and tested:

- Specimen A, finer cracks: crack widths in the SHCC in the range of 50–100 μm, and in the substrate 110–500 μm at the cracked face;
- Specimen B, wider cracks: crack widths in the SHCC in the range of 50–250 μm, and in the substrate 60–900 μm.

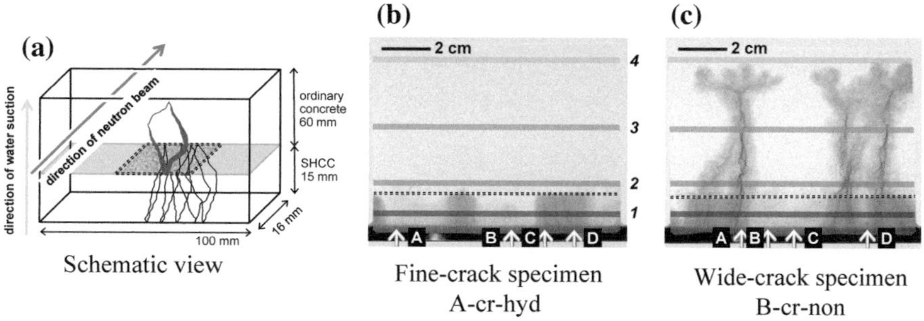

Fig. 8.16 **a** Schematic view of a cracked SHCC overlay on pre-cracked R/C substrate, with **b** finely cracked specimen A treated with water-repellent agent, and **c** wider cracked specimen B, without hydrophobic treatment (Schröfl et al. 2015)

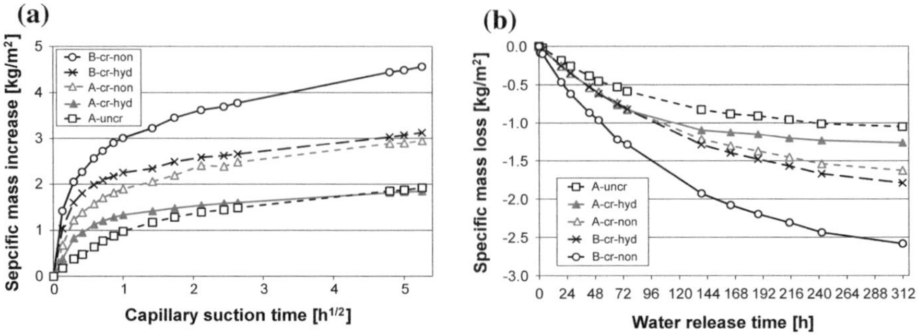

Fig. 8.17 Results of gravimetric measurement during **a** capillary absorption, and during **b** drying (Schröfl et al. 2015)

For each specimen type (A and B), two methods of substrate preparation were followed prior to application of the SHCC overlay, (i) without any form of water-repellent treatment, and depicted as specimen A-cr-non and B-cr-non, and (ii) hydrophobisation of the substrate by means of applying two brush-applied layers of commercial agent, depicted as A-cr-hyd and B-cr-hyd. The capillary water absorption was monitored by means of neutron radiography (for details see Schröfl et al. 2015).

Figure 8.16b, c shows the water distribution in specimens A-cr-hyd and B-cr-non, respectively, occurring, in both cases, after 32 mins of capillary absorption exposure. It is apparent that, for wide cracks in the SHCC overlay (specimen B), the cracked SHCC does not hinder ingress deep into the substrate. However, the combination of finer crack widths and of hydrophobisation prevents water absorption into the cracked substrate of specimen A-cr-hyd. Results of gravimetric measurement of water ingress through capillary absorption in the same test series are shown in Fig. 8.17a, and in Fig. 8.17b for drying. The fine crack widths in specimen A-cr-non appear to limit water ingress significantly compared with that in B-cr-non. However, ingress into the substrate is not completely prevented, with the ingress involved being subject to further investigation.

The increased resistance of cracked SHCC to fast capillary absorption, when a water-repellent agent was applied, was also demonstrated by Wang et al. (2014).

8.7 Recommendations for Durability Design and Conclusions

Several experimental and numerical investigations reported on in this chapter have shown that SHCC is a suitable material for repair layers on concrete substrate. The main reasons for its suitability are the crack-bridging capability of SHCC overlays and the higher resistance to drying shrinkage when compared to unreinforced repair mortar. In addition to the SHCC material properties, the interface behaviour could be demonstrated to have a significant influence on the fracture process. The latter is characterised by debonding in the interface and cracking in the SHCC overlay. By varying the interface roughness and bond strength, it is possible to balance these two fracture mechanisms (i.e. debonding and SHCC cracking), and to ensure both a monolithic mechanical behaviour of the structure and sufficiently small crack widths on the surface of the repair layer, allowing durability requirements to be met. Note that in the case of mechanical loading, a rather weak bond may be beneficial, as the crack widths in the SHCC tend to be smaller. In contrast, weak bonding is not recommendable when differential shrinkage is expected. The resulting comparatively large debonded interface area may have an unfavourable effect on the fatigue resistance, on the permeability, and on the resistance to fluid pressure within the cracks.

The tailoring of SHCC overlay systems for a certain application requires computational models that enable the prediction of the failure process. Mesoscale models, like lattice models, are effective tools for studying individual effects on the failure process, but they can hardly be used for structural design. For this reason, macroscale interface models have been proposed and the associated model parameters were identified by simulating laboratory experiments. Further validation of these models is still required.

For increasing the resistance to fast capillary absorption, hydrophobisation of the SHCC overlay by using water-repellent agents has proved to be an adequate technical solution.

References

Carol, I., Prat, P.C., López, C.M. (1997). Normal/shear cracking model: Application to discrete crack analysis. Journal of Engineering Mechanics 123(8):765-773.

Kamada, T., Li, V.C. (2000). The effects of surface preparation on the fracture behaviour of ECC/concrete repair system. Cement & Concrete Composites 22(6):423-431.

Kunieda, M., Kamada, T., Rokugo, K., Bolander, J.E. (2004). Localized fracture of repair material in patch repair systems. V.C. Li, C.K.Y. Leung, K.J. Willam, S.L. Billington (eds.), Fracture Mechanics of Concrete and Concrete Structures, Proceedings of the 5th International Conference on Fracture Mechanics of Concrete and Concrete Structures (FraMCoS–5), 12-16 April 2004, Vail, Colorado, USA, IA-FraMCoS, Volume 2, pp. 765-772.

Lim, Y.M., Li, V.C. (1997). Durable repair of aged infrastructures using trapping mechanism of Engineered Cementitious Composites. Cement & Concrete Composites 19(4):373-385.

Luković, M., Schlangen, E., Ye, G., Šavija, B. (2013). Impact of surface roughness on the debonding mechanism in concrete repairs. J.G.M. van Mier, G. Ruiz, C. Andrade, R.C. Yu, X. Zhang (eds.), Fracture Mechanics of Concrete and Concrete Structures, Proceedings of the 8th International Conference on Fracture Mechanics of Concrete and Concrete Structures (FraMCoS–8), 11-14 March 2013, Toledo, Spain, pp. 611-621.

Luković, M., Šavija, B., Schlangen, E., Ye, G., Van Breugel, J. (2014a). A modelling study of drying shrinkage damage in concrete repair systems. M.C. Forde (ed.), Proceedings of Structural Faults & Repair, 8-10 July 2014, London, UK.

Luković, M., Dong, H., Šavija, B., Schlangen, E., Ye, G., Van Breugel, K. (2014b). Tailoring strain-hardening cementitious composite repair systems through numerical experimentation. Cement & Concrete Composites 53:200-213.

Luković, M., Šavija, B., Schlangen, E., Ye, G. (2014c). Damage induced by continued corrosion in concrete repair systems. E. Schlangen, M.G. Sierra Beltran, M. Luković, G. Ye (eds.), Proceedings of the 3rd International RILEM Conference on Strain Hardening Cementitious Composites, 3-5 November 2014, Dordrecht, The Netherlands, RILEM Publications S.A.R.L., Bagneux, France, pp. 269-277.

Luković, M. (2016). Influence of interface and strain hardening cementitious composite (SHCC) properties on the performance of concrete repairs. PhD thesis, Delft University of Technology, The Netherlands.

Schröfl, C., Mechtcherine, V., Kaestner, A., Vontobel, P., Hovind, J., Lehmann, E. (2015). Transport of water through strain-hardening cement-based composite (SHCC) applied on top of cracked reinforced concrete slabs with and without hydrophobization of cracks – Investigation by neutron radiography. Construction and Building Materials 76:70-86.

Silfwerbrand, J., Beushausen, H., Courard, L. (2011). Bond. In: B. Bissonnette, L. Courard, D.W. Fowler, J.-L. Granju eds., Bonded Cement-Based Material Overlays for the Repair, the Lining or the Strengthening of Slabs or Pavements. State-of-the-Art Report of the RILEM Technical Committee 193-RLS, pp. 51-79.

Van Zijl, G.P.A.G., Stander, H. (2009). SHCC repair overlays for RC: Interfacial bond characterization and modelling. M.G. Alexander, H.-D. Beushausen, F. Dehn, P. Moyo (eds.), Proceedings of the 2nd International Conference on Concrete Repair, Rehabilitation and Retrofitting (ICCRRR-2), 24-26 November 2008, Cape Town, South Africa, Taylor & Francis Group, London, UK, 2009, pp. 995-1003.

Wagner, C., Slowik, V., Waldenburger, K. (2008). Dehnungsverfestigendes zementgebundenes Material für die Sanierung gerissener Betonflächen (Strain-hardening cement-based material for the repair of cracked concrete surfaces). Bautechnik 85(1):49-56.

Wagner, C., Slowik, V. (2011). Strain hardening cement-based composites for repair layers on cracked concrete surfaces. M. Grantham, V. Mechtcherine, U. Schneck (eds.), Proceedings of Concrete Solutions, 4th International Conference on Concrete Repair, 26-28 September 2011, Dresden, Germany, Taylor & Francis Group, London, UK, 2012, pp. 775-782.

Wagner, C., Bretschneider, N., Slowik, V. (2013). Characterization of the interface between strain hardening cementitious repair layers and concrete subgrade. J.G.M. van Mier, G. Ruiz, C. Andrade, R.C. Yu, X. Zhang (eds.), Fracture Mechanics of Concrete and Concrete Structures, Proceedings of the 8th International Conference on Fracture Mechanics of Concrete and Concrete Structures (FraMCoS-8), 11-14 March 2013, Toledo, Spain, pp. 2045-2055.

Wagner, C., Bretschneider, N., Villmann, B., Slowik, V. (2014). Modelling of the bond between strain hardening cementitious repair layers and concrete substrate. E. Schlangen, M.G. Sierra Beltran, M. Luković, G. Ye (eds.), Proceedings of the 3rd International RILEM Conference on Strain Hardening Cementitious Composites, 3-5 November 2014, Dordrecht, The Netherlands, RILEM Publications S.A.R.L., Bagneux, France, pp. 279-286.

Wagner, C. (2016). Dauerhaftigkeitsrelevante Eigenschaften von dehnungsverfestigenden zementgebundenen Reparaturschichten auf gerissenen Betonuntergründen (Durability-related properties of strain-hardening cement-based repair layers on cracked concrete substrates). Doctoral thesis, Technische Universität Dresden, Germany.

Wang, P., Wittmann, F.H., Zhang, P., Lehmann, E., Zhao, T. (2014). Durability and service life of elements made with SHCC under imposed strain. E. Schlangen, M.G. Sierra Beltran, M. Luković, G. Ye (eds.), Proceedings of the 3rd International RILEM Conference on Strain Hardening Cementitious Composites, 3-5 November 2014, Dordrecht, The Netherlands, RILEM Publications S.A.R.L., Bagneux, France, pp. 33-41.

Zhang, Y.X., Ueda, N., Umeda, Y., Nakamura, H., Kunieda, M. (2011). Evaluation of shear failure of strain hardening cementitious composite beams. Procedia Engineering 14:2048-2057.

Chapter 9
Reinforcing Bar Corrosion

Koichi Kobayashi, Suvash C. Paul and Gideon P.A.G. van Zijl

Abstract Currently, a common use for strain-hardening cement-based composites (SHCC) is as a repair material, or for retrofitting reinforced concrete (RC) structures. This is due to SHCC being expected to have high resistance to substance penetration as the cracks that are produced in SHCC are fine. Equally, in retrofitting applications such as those described in Sects. 1.5.3–1.5.5, or in applications of steel-reinforced SHCC (R/SHCC) in structures in coastal regions, the corrosion of the steel reinforcement is likely to determine both durability and structural service life. Accordingly, this chapter discusses the chloride-induced corrosion of R/SHCC. From reported experimental results, chloride profiles in cracked R/SHCC have been reported in Chap. 2. Here, the chloride contents are evaluated for correlations with observed corrosion damage in the steel bars for different cover depths in cracked R/SHCC. Steel bar damage is expressed in terms of steel mass loss, corrosion depth, and reduction in yield resistance. Finally, a corrosion model is proposed for R/SHCC, incorporating crack width, crack spacing, free chloride content, and cover depth.

Keywords Chloride · Crack width · Crack spacing · Corrosion · Pitting

9.1 Introduction

As outlined in Chap. 1, strain-hardening cement-based composites (SHCC) exhibit strain-hardening behaviour in tension, due to the formation of multiple fine cracks, upon increased deformation. The width of these cracks is relatively small, typically

---

K. Kobayashi (✉)
Gifu University, Gifu, Japan
e-mail: ko2ba@gifu-u.ac.jp

S.C. Paul
Nanyang Technological University, Singapore, Singapore

G.P.A.G. van Zijl
Stellenbosch University, Stellenbosch, South Africa

G.P.A.G. van Zijl and V. Slowik (eds.), *A Framework for Durability Design with Strain-Hardening Cement-Based Composites (SHCC)*, RILEM State-of-the-Art Reports 22,
DOI 10.1007/978-94-024-1013-6_9

less than 0.1 mm, which is potentially beneficial for resistance to the ingress of water, water-soluble substances, and gases, therefore enabling high durability. For this reason, SHCC and steel-reinforced SHCC (R/SHCC) have been used to conduct repairs or retrofitting in corrosion-deteriorated reinforced concrete (RC) structures. The reader is referred to Sects. 1.5.3–1.5.5 for applications of R/SHCC retrofitting in RC structures exposed to chloride-induced corrosion either in coastal regions or in regions of salt-based de-icing practise. In these applications, subsequent structural life depends on the corrosion resistance of R/SHCC, despite chloride exposure and crack formation under imposed strain. Another repair strategy could be to apply SHCC as an overlay onto a RC substrate to strengthen the structural system, as well as so as to protect any reinforcing steel in the substrate from corrosion, as was alluded to in Sect. 8.6 in the previous chapter. In this chapter, the focus is on the corrosion of steel-reinforcing bars in R/SHCC.

Until now, only limited SHCC-related data have emerged from both field and laboratory experimental testing. Currently, no corrosion model exists for this new type of construction material, with the existing corrosion models for conventional RC also being open to criticism due to the fact that they sometimes either overestimate or underestimate the actual amount of corrosion occurring (Otieno et al. 2010).

Corrosion is a complex deterioration mechanism and there is no direct way of determining the actual corrosion rate, which complicates corrosion modelling and model parameter characterisation. In this chapter, non-destructive methods are used to measure the corrosion rate, and final verification is based on destructive methods to remove the bars for the evaluation of corrosion deterioration after a particular period of exposure. So far, a limited set of parameters has been studied, including the influence of cover depth, crack spacing, and exposure duration. To a lesser extent, matrix types have been studied, in the form of a low fly ash content versus high fly ash content binder, and fine sand versus coarse sand.

Two fibre types, namely polyvinyl alcohol (PVA) and high-modulus polyethylene (PE) fibres, are considered here, based on reported experimental results in the literature. Typical chloride exposures have been chosen as cyclic wetting–drying with chloride solutions in water, which are believed not to alter the electrochemical corrosion cell expected in typical field applications in coastal regions and/or in highway infrastructure subjected to de-icing salting practices. Comparative results of R/SHCC and R/mortar subjected to high electrical potential are reported, indicating the higher corrosion resistance of R/SHCC when compared to R/mortar under these particular exposure conditions.

In RC, a focus on increasing service life relies on providing good quality, low-permeable concrete and sufficient cover depth to reinforcing steel, so as to delay the point of corrosion initiation, or to increase the so-called corrosion initiation period. This remains a scenario requiring consideration in the durability design of R/SHCC, namely the eventual depassivation of reinforcing steel in uncracked R/SHCC through the gradual ingress of chloride, oxygen and water.

However, another scenario should be considered in durability design, where crack formation due to imposed strain allows the quick ingress of deleterious substances to the surface of the steel in R/SHCC. Under these circumstances, the initiation period may be neglected, whereas the low corrosion rate in finely cracked R/SHCC is relied upon for a suitable structural service life. The low corrosion rates in cracked R/SHCC, as reported by Miyazato and Hiraishi (2005), indicate the potential of the latter strategy. In the subsequent sections of this chapter, corrosion in R/SHCC under chloride exposure is discussed. The topics of corrosion depth, corrosion area, mass loss, and the loss of yield force of steel bars are discussed and related to the crack patterns in the R/SHCC specimens. Finally, based on these parameters, corrosion modelling is also proposed for single and multiple cracks in R/SHCC specimens.

9.2 Chloride Ingress into Cracked SHCC

In Sect. 2.5, ingress of chloride into cracked R/SHCC was discussed. The results of several experimental series of the exposure of cracked R/SHCC specimens to chloride are presented in the section mentioned. Methods of chloride profiling are elaborated, in terms of XRF (X-ray fluorescence), as well as in terms of chemical titration. The Ingress Potential Index (IPI_{Cl}) and the crack width value (*CWV*) developed by Boshoff et al. (2016) and Wagner (2016) respectively, and which are described in Sect. 2.2, are calculated for specific experimental specimens. The correlation between the indices in question and the total chloride content in cracked SHCC specimens is shown to be reasonable for relatively short periods of exposure, but it is regarded as being less reasonable for long periods of cyclic exposure (see Fig. 2.20). The pool of data is acknowledged not only as being limited at present, but also as requiring extension in future work. Also, the IPI_{Cl} and *CWV* indices do not incorporate the mechanism of chloride ion transfer in non-saturated, cracked SHCC, for which further research is recommended.

Whereas a reasonable correlation has been shown between the indices and chloride content, it still remains to be seen whether there is a correlation with actual corrosion damage in R/SHCC.

9.3 Chloride-Induced Corrosion of Steel Bars in SHCC

Non-destructive ways of measuring corrosion rate have been proposed for RC. Such methods might offer non-destructive ways of presenting the reinforcement condition in R/SHCC. The different methods that can be used to determine the corrosion rates of specimens are influenced by various parameters, including temperature, humidity, and the presence of moisture and oxygen in the specimens concerned. Notably, a corrosion rate reading depends on the parameters and conditions prevailing at the time when the measurement is taken. Therefore, the corrosion rate value cannot be expected to remain consistent, since the aforementioned parameters may vary from day to day, for which compensation must be made. Also, various readings over time are required so as to be able to estimate the actual state of corrosion damage. Due to corrosion, the cross-section of a steel bar reduces either uniformly, or in localised, pitting fashion, resulting in a reduction of its tensile resistance capacity. Paul and Van Zijl (2014) proposed Eq. 9.1 for estimating the uniform corrosion depth of the steel bar (d_c) when the corrosion rate of a steel bar, over time, is known. Since mass loss and tensile resistance/force capacity loss are related to the section loss of steel bars, Eq. 9.1 allows for the estimation of its mass and resistance loss due to corrosion in cases of uniform corrosion:

$$d_c(t) = d_{c0} + \int_0^t V_{corr} dt = d_{c0} + \sum_{i=1}^{N_t} \frac{1}{2} \left(V_{corr} + V_{corr,i-1}\right)(t_i - t_{i-1}) \tag{9.1}$$

with:

d_{c0} the initial corrosion depth
V_{corr} the corrosion rate (mm/year)
t_i the period of corrosion

9.3.1 Rate of Corrosion of Steel Bars in R/SHCC

The corrosion rate in SHCC class materials has been examined by several researchers (Şahmaran et al. 2007, 2008; Kobayashi et al. 2010; Mihashi et al. 2011; Paul 2015; Jen and Ostertag 2016) over the last few years. All the authors mentioned have reported the superior performance of SHCC materials over standard repair, and other mortars, as well as over conventional concrete. However, different methodologies for corrosion measurement were used by the different researchers, and the link to the specific exposure and deterioration process has not yet been made clear in all cases.

9.3.1.1 Uncracked Lollypop Specimens

Şahmaran et al. (2008) exposed SHCC and mortar lollypop-shaped specimens with centrally embedded steel bars to 5% NaCl solution, applying a potential of about 30 V to the steel bar. The steel bars had an SHCC/mortar covering of approximately 30 mm. Crack formation in the mortar and SHCC specimens, due to the corrosion of the steel bars, was monitored. After 300 h of testing, the corrosion current in the SHCC specimen was below 1 A, whereas the corrosion current was more than 1.5 A after 75 h of testing in the case of the mortar specimen (see Fig. 9.1). A single large crack of about 2 mm wide was observed to have formed in the mortar specimen parallel to the steel bar, whereas multiple fine cracks, of about

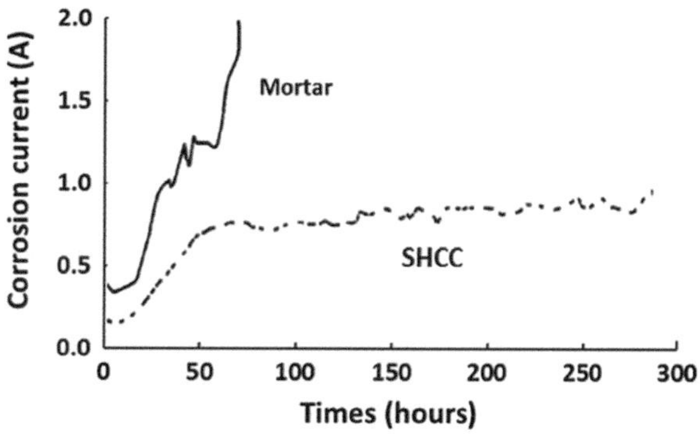

Fig. 9.1 Observed corrosion currents in R/SHCC and R/mortar (Şahmaran et al. 2008)

0.01 mm in width, were found in the SHCC specimens after the same accelerated corrosion testing had taken place. Almost 12% mass loss of steel was observed in the mortar specimen after 75 h of the accelerated corrosion testing, whereas no, or insignificant, mass loss of the steel bar was noticed in the case of the SHCC specimen. After 300 h of accelerated testing, mass loss of 18% was found in the steel bar in the SHCC specimen. The researchers concerned concluded that, if a 0.3 mm wide crack width is considered to be a serviceability limit state for a structure, the service life of R/SHCC would be 15 times as high as that of R/mortar concrete structures.

9.3.1.2 Cracked Specimens, w/c = 0.3, Polarisation Method

Miyazato and Hiraishi (2013) investigated the corrosion rate in flexurally cracked high-performance fibre-reinforced cementitious composite (HPFRCC) and mortar specimens with w/c ratios of 0.3 and 0.6. Specimens were exposed to 3% NaCl solution, with penetration being allowed only through the cracked face of the specimen by means of capillary absorption, as the other faces of the specimen were sealed with epoxy resin. Such exposure was maintained for two days, followed by drying for five days, with the same cycle being followed for 28 days. Using the polarisation resistance corrosion measurement technique, for w/c of 0.3, corrosion rates of 0.082 and 0.004 mm/year were found respectively in R/mortar and R/HPFRCC specimens. The reason for the lower corrosion rate of the latter material was understandable in terms of the shallower crack depth and the smaller crack width of such specimens in comparison to those of the mortar specimens. Also, the multiple small cracks in the R/SHCC, by means of which chloride had penetrated to the level of the steel in various positions, enabled the formation of microcell corrosion, rather than the macrocell corrosion that had occurred in the R/mortar at the location of the single large crack.

9.3.1.3 Macrocell Versus Microcell Corrosion

In the above-mentioned research undertaken by Miyazato and Hiraishi (2013), the corrosion rates in the R/SHCC type specimens were found to be significantly (15–20 times) lower than they were in the R/mortar specimens. Insignificant corrosion was also observed in the case of steel reinforced ductile-fibre-reinforced cementitious composite (R/DFRCC) beams containing multiple cracks, compared to the amount of corrosion that was observed in cracked RC beams by Maalej et al. (2003). Significant macrocell corrosion (according to the lower anode to cathode area ratio) was found in singly cracked R/mortar in comparison to the insignificant corrosion that was found to occur in multiply cracked R/SHCC specimens, due to the formation of dominantly occurring microcell corrosion (according to the higher anode to cathode area ratio) by Miyazato and Hiraishi (2005). In terms of electrochemical corrosion processes, relatively high anode to cathode area ratios betoken a relatively high resistance of ion movement. As a result of such resistance, the corrosion damage that is due to microcell formation in steel bars tends to be relatively low. However, in contrast, in terms of macrocell corrosion, a large cathode area is paired with a small area of anode, and the ion movement from the cathode to the anode is high. The high ion movement in macrocell corrosion leads to the formation of localised pitting corrosion damage at the anode, which significantly reduces the cross-section of the steel bar. For example, it is believed that the corrosion damage in a steel bar, due to the presence of a corrosion rate of 0.5 mm/year in a relatively small anode area, is much higher than is the same corrosion rate in a larger anode area (Song and Shayan 1998).

9.3.1.4 Cracked Specimens with Different Cover Depth

The corrosion rate in R/SHCC specimens with different cover depths (15, 25, and 35 mm, denoted by C15, C25, and C35) and a single reinforcing steel bar (B1) was measured by means of a Coulostatic method, as is shown in Fig. 9.2, by Paul and Van Zijl (2014). Specimens, which were stored in the laboratory at ambient temperature, were subjected to accelerated chloride exposure in the form of cyclic wetting (3 days), with chloride solution in water (3.5% NaCl), and drying (4 days). The corrosion rate reading was taken once a week. Wiggles in the corrosion rate readings are evident in Fig. 9.2, the presence of which was ascribed to sensitivity to variations in the temperature, humidity, and moisture content of the specimens. However, upon integration, according to Eq. 9.1, relatively smooth corrosion depth (d_c) values, as shown in Fig. 9.3, were found, allowing for improved interpretation of steel bar corrosion in the various specimens tested.

Corrosion depths that were calculated in the above-mentioned way were found to be relatively high for small steel bar cover depths. In the specimens concerned, the total chloride content at the steel surface level was found, generally, to be relatively high (see Fig. 2.16). The relationship between free and total chloride

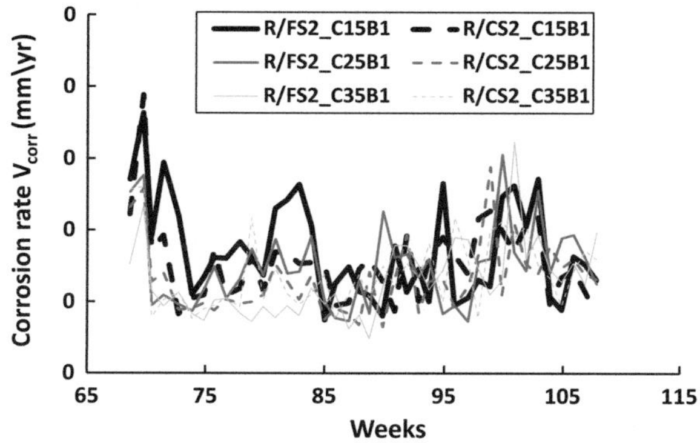

Fig. 9.2 Corrosion rate in different cover depths of R/FS2 (fine sand) and R/CS2 (coarse sand) specimens with single steel bars (B1) (Paul and Van Zijl 2014)

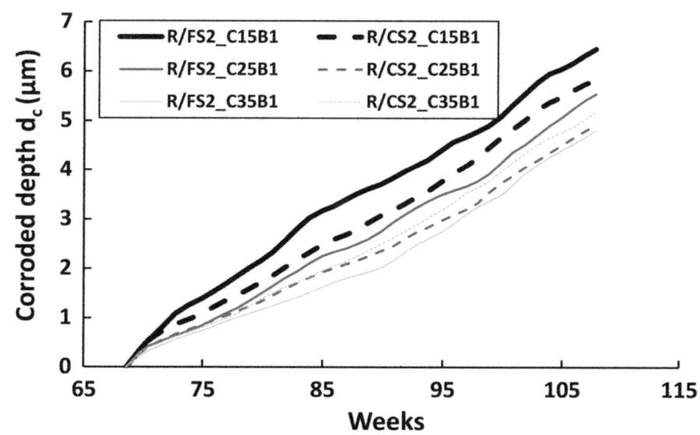

Fig. 9.3 Estimated corrosion depths in different cover depths of R/FS2 (fine sand) and R/CS2 (coarse sand) specimens with single steel bars (B1) (Paul and Van Zijl 2014)

derived from water-soluble and acid-soluble chemical titration testing, as given in Fig. 2.18, may be relevant for the particular SHCC matrix type used.

Another advantage of SHCC, in terms of preventing rebar corrosion, is the large unit mass of the binder in its mixture. This results in a great capacity for binding/fixing chloride, whereby a relatively low amount of free chloride that damages the passive film on the rebar, is available.

The correlation between cover depth, crack characteristics, and corrosion depth is discussed in the next section (see Figs. 9.2 and 9.3). The correlations in Sect. 9.3.2 were validated by means of actual inspection and measurement of the steel bars taken from specimens. Pitting depths, the loss of yield resistance, and the mass loss of steel bars reflect such corrosion damage (see Sect. 9.3.3).

9.3.2 Influence of Cover Depth, Cracks and Chloride on Corrosion in R/SHCC

Figure 9.4 shows the influence of cover depth on corrosion depth in R/SHCC specimens made from fine sand (R/FS2, with a maximum sand particle size of 0.30 mm) and coarse sand (R/CS2, with a maximum sand particle size of 1.70 mm) under conditions of chloride capillary absorption exposure for precracked specimens in the unloaded condition (Series 1, FS2), as was described in Sect. 2.5. The materials composition was reported on in Table 2.2, with such crack properties as the number of cracks, average and maximum crack widths ($W_{cr,ave}$ and $W_{cr,max}$) in a gauge length of 200 mm of a total of 500-mm-length R/SHCC specimens under conditions of flexural testing are also shown in Fig. 9.4. Corrosion depths were found to be relatively low for the comparatively high cover depth of steel bars, with an apparent threshold of between 25 and 35 mm cover depth in the particular case under discussion.

The influence of average crack width and average crack spacing on corrosion depths in the R/FS31 specimens (Series 2 by Paul (2015) in Table 2.2) is shown in Fig. 9.5. The corrosion depth was found, generally, to be relatively large for comparatively large average crack widths, and for relatively large crack spacing. The possibility that the average crack width and spacing were most likely not independent warrants further investigation. Finally, the corrosion depths versus total and free chloride contents at the surface of steel bars in the different R/SHCC specimens are shown in Fig. 9.6. Free chloride content was found to be relatively strongly correlated to the corrosion involved.

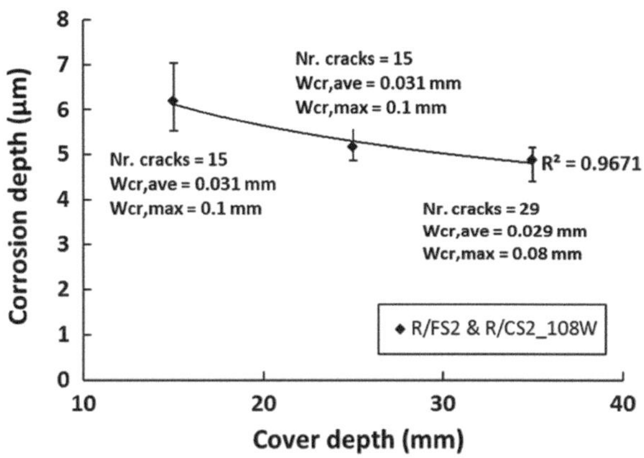

Fig. 9.4 Relationship between d_c (Eq. 9.1) and cover depth (Paul 2015)

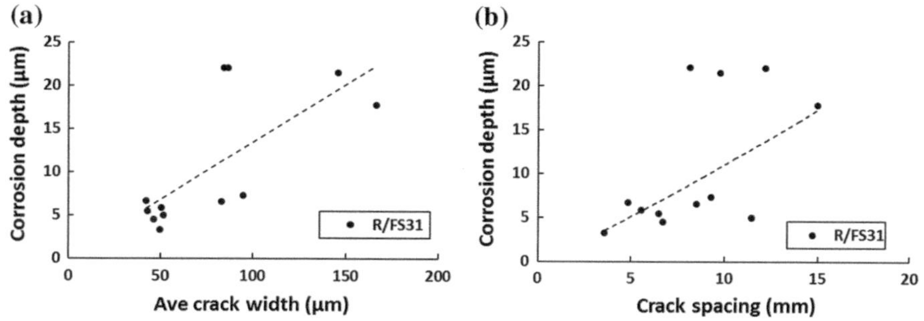

Fig. 9.5 Influence of **a** average crack widths and **b** crack spacing on corrosion depth in R/SHCC specimens, after 57 weeks of exposure (Paul 2015; Paul and Van Zijl 2016)

Fig. 9.6 Relationship between corrosion depth and chloride content (Paul 2015; Paul and Van Zijl 2016)

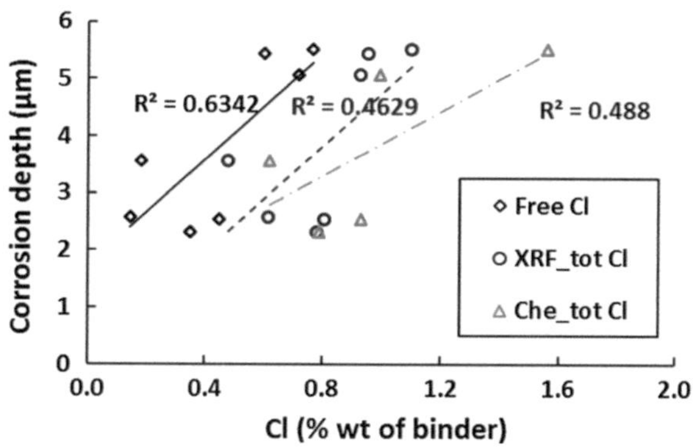

9.3.3 Progress of Steel Bar Corrosion Deterioration

In the case of ordinary concrete, the intrusion of water and of deleterious ions that cause deterioration is increased by the occurrence of corrosion cracking in the cover concrete, which accelerates the corrosion concerned. However, in SHCC, even if corrosion does occur, the subsequent expansion and corrosion cracking is restrained, so the rate of progress of the deterioration tends to be relatively insignificant. The current section describes corrosion damage in the steel bar.

Subsequent to the non-destructive testing of corrosion rates by Paul (2015), which has been described in Sects. 9.3.1 and 9.3.2, steel bars, on their removal from the specimens involved, were inspected for actual corrosion damage. In addition, other researchers have performed destructive testing, so as to be able to characterise the actual corrosion damage, as is reported in the following subsections.

9.3.3.1 Corrosion Area in Steel Bar

Kobayashi et al. (2010) used HPFRCC containing three different fibre volume percentages as a surface repair coating (above steel reinforcement) as well as as a patch repair material stretching behind the steel reinforcement (see Fig. 9.7) of RC specimens. The corrosion protection performance of the repairs concerned was observed. Cracks were formed in the specimens by means of tensile testing, after which 3% NaCl solution was sprayed over the specimens for five minutes every six hours for 60 days of accelerated corrosion testing. The steel bars were subsequently removed, and the corrosion area was expressed as a percentage of the total bar surface area. To determine the corrosion area involved, the outline of each bar was traced onto a plastic sheet, with the corrosion area then being measured in relation to the discoloured area, by means of the use of a planimeter (see Fig. 9.8 for the corrosion areas). A minimum of 2% to a maximum of 10% corrosion area was

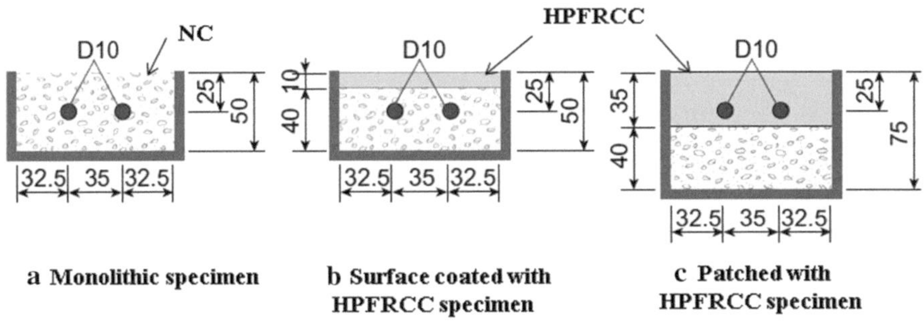

Fig. 9.7 Applied HPFRCC in RC specimens (Kobayashi et al. 2010)

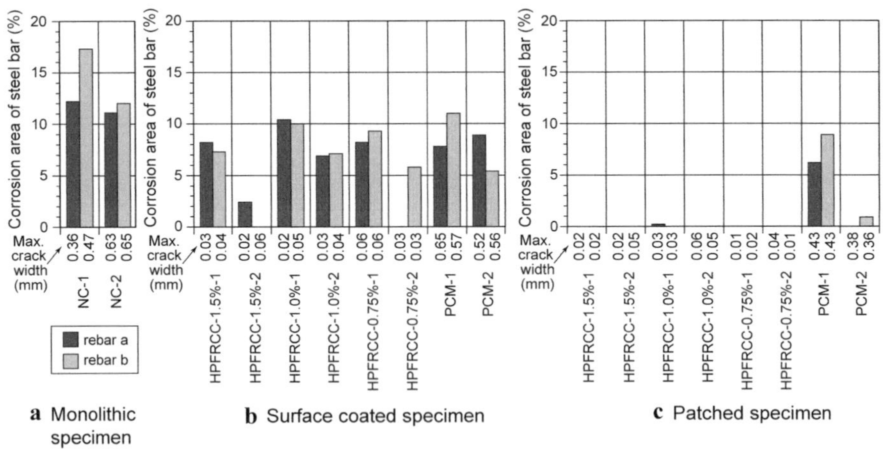

Fig. 9.8 Corrosion areas on the two steel rebars in each specimen (Kobayashi et al. 2010)

observed in all the steel bars that were present in the surface-coated specimens. However, in the patch repair specimens, significantly lower corrosion area was found in the steel bars than was found in the other specimens. The significance of the decrease concerned was ascribed to the relatively fine crack widths that formed in the HPFRCC patch repair. No significant difference was found in the performance of the HPFRCC with three different fibre contents.

Figures 9.9 and 9.10 shows the crack patterns of the R/SHCC and R/mortar specimens and corroded bars that were removed from R/SHCC and R/mortar specimens (given in Table 2.2 by Paul (2015)). Note that the mortar specimens, although having the same matrix as did the SHCC specimens, contained no fibre. Both of the bars concerned are, firstly, shown uncleaned, and then after cleaning

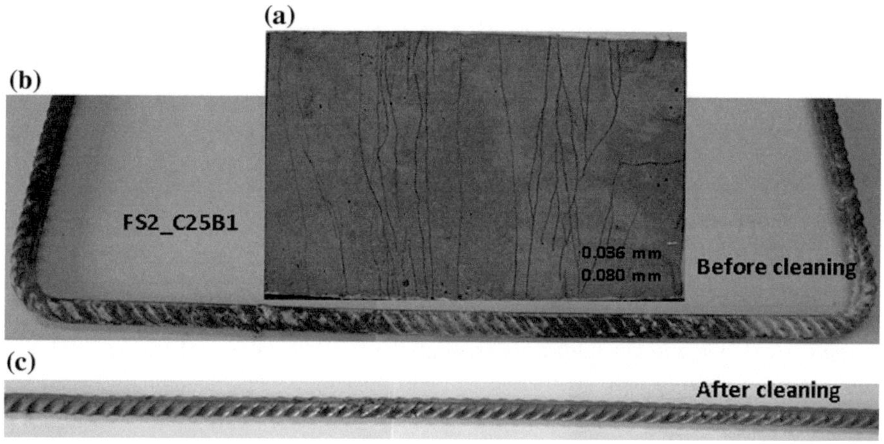

Fig. 9.9 Inspection of corrosion damage in an R/SHCC specimen (Paul 2015)

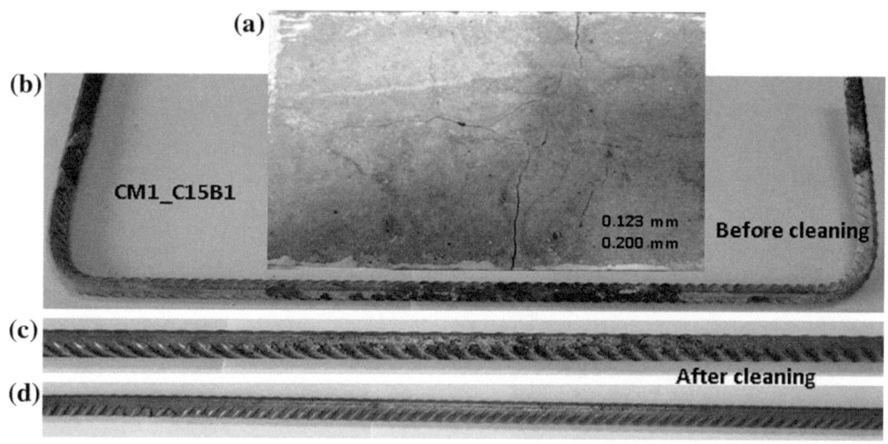

Fig. 9.10 Inspection of corrosion damage in an R/mortar specimen (Paul 2015)

with hydrochloric acid (HCl) acid. The average and maximum crack widths in the specimens are also indicated in the relevant figures. The distributed corrosion stains that were found over the entire length of the steel bar were ascribed to the presence of multiple fine cracks in the R/SHCC specimens. An increased number of corrosion activities were seen clearly outside the cracked region of the R/SHCC specimens, whereas, in the R/mortar specimens, the corrosion was more localised around the wider crack widths. The reason for the distributed corrosion in R/SHCC outside the cracked region are explained as follows: (i) The levels of deformation in the R/SHCC specimens in flexural testing were relatively high, so that delamination is likely to have occurred in the matrix and the steel surface, creating a path through which chloride could travel along the steel bar length; (ii) The parallel splitting and cracking that was also observed in some of the R/SHCC specimens enabled chloride penetration to take place parallel to the steel bars. Several cases of localised pitting corrosion areas were observed in the steel bars, especially in the cracked region of R/mortar specimens. In the R/SHCC specimens, distributed pitting was observed in the cracked region, as well as outside the cracked region. The detailed characterisation of pitting corrosion is discussed in Sect. 9.3.3.3.

9.3.3.2 Corrosion Mass Loss

In applying cyclic chloride exposure to uncracked hybrid fibre reinforced cementitious composites (HFRCC) and mortar beam specimens, each of which contained a single steel bar and an additional potential of 3 V to the embedded steel bar, Mihashi et al. (2011) recorded the corrosion current present for up to 52 weeks. Using Faraday's law, mass loss was calculated in the HFRCC specimens, as well as in the mortar specimens. After cyclic wetting (during which the specimens' bottom surface was submersed in 3% NaCl solution for 3.5 days) and drying (also for 3.5 days) for a year, only 10 g of mass loss was found in the HFRCC specimens, whereas, in the mortar specimens, mass loss exceeded 50 g, as is shown in

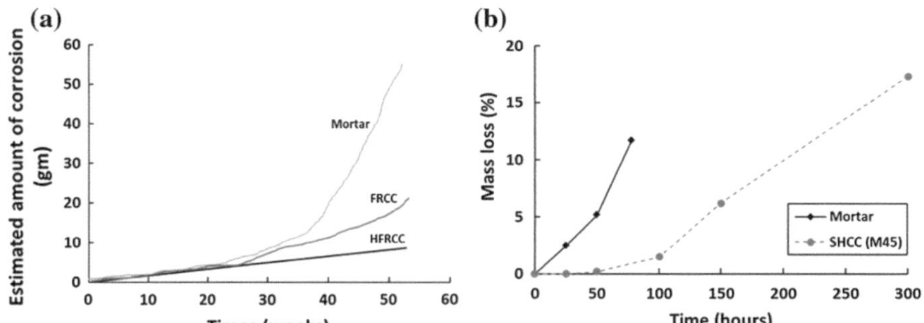

Fig. 9.11 Mass loss versus corrosion exposure time for SHCC and mortar corrosion specimens, as reported by **a** Mihashi et al. (2011), and by **b** Şahmaran et al. (2007)

Fig. 9.11a. Also, 100% of the steel area was affected by corrosion in the mortar specimens, and a corrosion depth of about 3.1 mm was found in the original 13-mm-diameter polished rebar. In the HFRCC, only 65% of the steel area was affected by corrosion, with the depth of the corrosion being limited to 1.2 mm. The authors in question concluded that the superior crack-bridging and self-healing capacity of HFRCC may have been the reason for the relatively low mass loss in the case of the steel that was embedded in the HFRCC in comparison to the mass loss that was experienced by the steel that was embedded in the mortar. Note that the mortar and HFRCC consisted of different mix designs, unlike was the case with the testing in the case of Paul (2015).

Figure 9.11 shows the results that were achieved by Mihashi et al. (2011), and those that were achieved by Şahmaran et al. (2007), which were reported in Sect. 9.3.1, in terms of mass loss evolution over time. Both of the test series concerned were performed by subjecting the specimens involved to additional potential. In both cases, the mass loss that occurred in the reinforcement bars in the SHCC/HFRCC specimens was significantly less than was that in the reinforcement bars in ordinary, or repair-type, mortars. Furthermore, in the lollypop mortar test specimens, after corrosion cracking had taken place during the period of the test, the progression of the corrosion involved could be seen to accelerate. In contrast, in the SHCC/HFRCC test specimens, no cracks (Mihashi et al. 2011), or only fine ones (of 0.01 mm, Şahmaran et al. 2007) appeared during the test period, with no acceleration of the corrosion being observed in the case of Mihashi et al. (2011). In the case of the finely cracked lollypop SHCC specimens, only slight acceleration was observed (Şahmaran et al. 2007).

The actual mass loss of the steel bars due to the presence of the different number of cracks and crack spacing in notched R/FS32 specimens (Series 3 in Table 2.2, Paul (2015)) are shown in Fig. 9.12. The specimens involved were similar to those in Series 1 and 2, but, instead of the natural flexural crack spacing in R/SHCC, notches were sawn into the flexural face of the specimens after curing, so as to cause a single central flexural crack (in the case of specimens denoted N1), three central cracks, spaced 40 mm apart (in the case of specimens denoted N3), and five

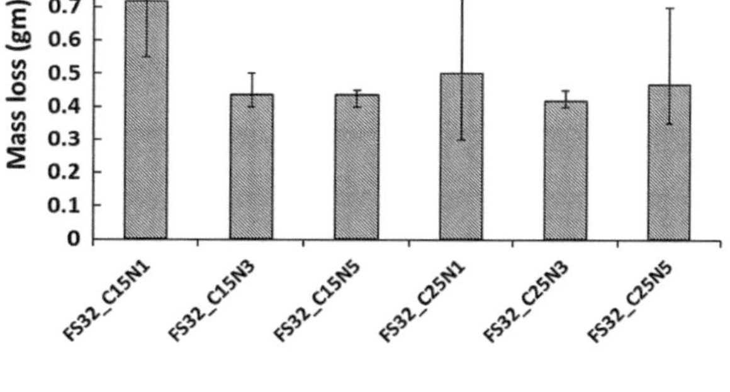

Fig. 9.12 Actual mass loss in the steel bars, due to corrosion in the R/FS32 specimens after 28 weeks of chloride exposure (Paul 2015), with C15 and C25 denoting 15 and 25 mm cover depth. N1, N3 and N5 denote 1, 3 and 5 notches, respectively

central cracks, spaced 20 mm apart (in the case of specimens denoted N5). Two cover depths were also considered (15 and 25 mm, which were denoted C15 and C25). Mass losses of steel bars was determined by means of measuring the weight difference before placing them in the moulds for casting, after which they were removed from the specimen, subsequent to cyclic chloride ponding exposure to the cracked face, exactly as was done in the case of Series 2. The procedure involved consisted of three days of wetting, followed by four days drying, in the loaded condition throughout, and within special steel frames, which, in the case discussed, lasted for a period of 28 weeks. Both before casting and after removal from the specimen, the bars were cleaned with HCl acid. Mass loss in the single-notch specimens was found to be higher than it was in the case of the other specimens. The heightened mass loss can be explained in terms of the increased corrosion rate in the relatively small anode area of the N1 specimens, which caused the heightening of loss in comparison to the lowered corrosion rate in the relatively large anode area specimens of N3 and N5. In R/FS32 specimens with a cover depth of 15 mm, the average mass loss in N1 specimens was 60 and 67% higher than it was in the specimens with three and five notches that were spaced 40 and 20 mm apart, respectively.

9.3.3.3 Pitting Corrosion Depth

Consequences of corrosion in terms of pitting depth and pitting area were also observed by Paul (2015) in R/SHCC specimens. The presence of chloride in concrete which absorbs water and retains moisture in the pores, results in an increase in the electrical conductivity in concrete. Relatively high conductivity of concrete contributes to the separation of anode and cathodes, as the ions concerned

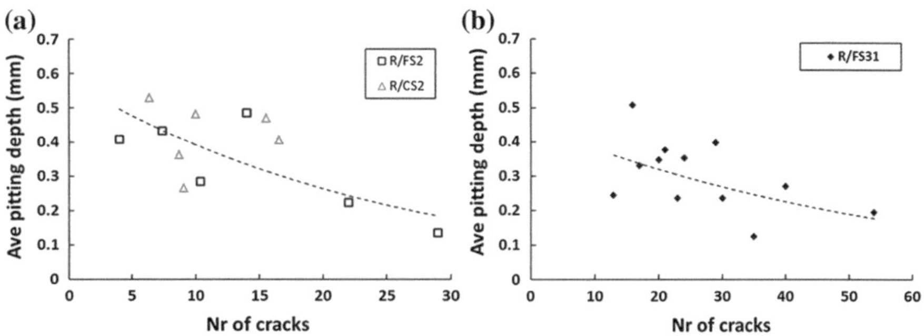

Fig. 9.13 Pitting depth with different numbers of cracks in the 200-mm-long exposed length of R/SHCC specimens of **a** Series 1, and **b** Series 2 in Table 2.2 (Paul 2015)

can move through the water-lined pores. Due to these ions, pitting can occur in the steel bar, which can significantly reduce the steel capacity.

The corrosion pitting depths in the individual steel bars were found to be relatively low for an increased number of cracks within the exposed length of 200 mm in the R/SHCC specimens, as is shown in Fig. 9.13. The phenomenon can be explained by the fact that an increased number of cracks in the specimen over a given length naturally produces comparatively fine average crack spacing, leading to relatively small cathodic areas in comparison to the anodic area in the corrosion cell. Accordingly, comparatively few OH^- ions become available from the relatively small cathode to react with the Fe^+ in the anode. In such case, the corrosion that is distributed over the lengths of steel bars is expected to have distributed pitting (see Fig. 9.9), rather than consisting of concentrated corrosion (see Fig. 9.10). Again, crack spacing and crack width are acknowledged as being somehow related, which warrants further investigation.

The correlation between the total crack width (being the sum of all crack widths in the 200-mm-long exposed surface) and the average crack spacing with the pitting corrosion depths was also studied. For total crack widths below 1 mm (in the gauge length of 200 mm of specimens tested in the current series), a trend of reduced pitting depths with increased total crack width is seen in Fig. 9.14a for Series 1 specimens, which were tested in the unloaded condition. For Series 2, which were tested in the loaded condition, relatively large total crack widths (>1 mm) occurred, with, for the present series of specimens, the trend not being as significant (see Fig. 9.14b). The trend of reduced pitting depths, with increased total crack width, may be ascribed to microcell corrosion in the steel bar, due to the presence of a relatively high number of cracks being associated with a comparatively high total crack width (see Fig. 9.15). However, the relatively high number of cracks is typically also often associated with the presence of an increased number of cracks of comparatively fine width in SHCC that is precracked by being bent to the same deflection. Thus, an increase in both the closeness of the crack spacing, and in the fineness of the crack width may have led to distributed corrosion.

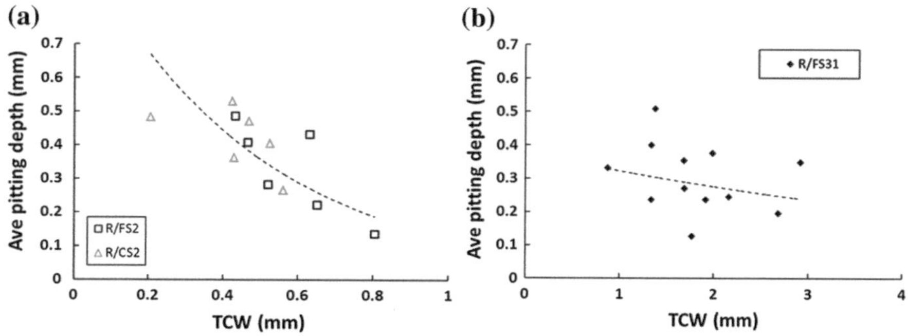

Fig. 9.14 Average pitting depth versus total crack widths (TCWs) in R/SHCC specimens of **a** Series 1, and **b** Series 2, in Table 2.2 (Paul 2015)

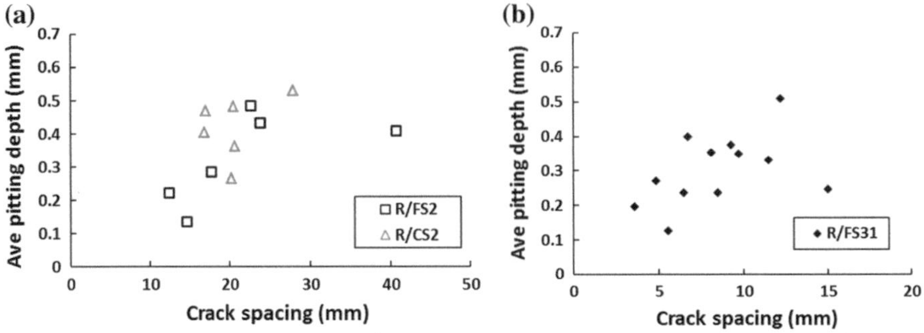

Fig. 9.15 Average pitting depth versus crack spacing in R/SHCC specimens of **a** Series 1, and **b** Series 2, in Table 2.2 (Paul 2015)

9.3.3.4 Localised Versus Distributed Corrosion Damage

Typically, in the steel bars studied, more severe pitting depths have been found due to macrocell corrosion than due to microcell corrosion (Broomfield 2007). The postulation was also verified by means of measuring and adding up the total pitting area in the steel bars, as is shown in Fig. 9.16 for both Series 1 and 2 in Table 2.2. The actions in question were performed after cleaning each steel bar with HCl, and then carefully marking each visible pitting area with appropriate simple surface geometrical shapes (circular, rectangular, and trapezoidal). A Vernier scale was then used to measure the marked areas accurately, so as to be able to calculate the total pitting area on the surface of each steel bar. From Fig. 9.16, it is apparent that a relatively large number of cracks in the specimens caused the lowering of the overall pitting area.

The influence of total crack widths on the steel-pitting diameter at the crack locations of the corrosion steel bars in RC was also examined by Mohammed et al. (2001). After corrosion testing had taken place, the RC specimen was broken, and the pitting diameter in the corroded steel bars was measured by means of a

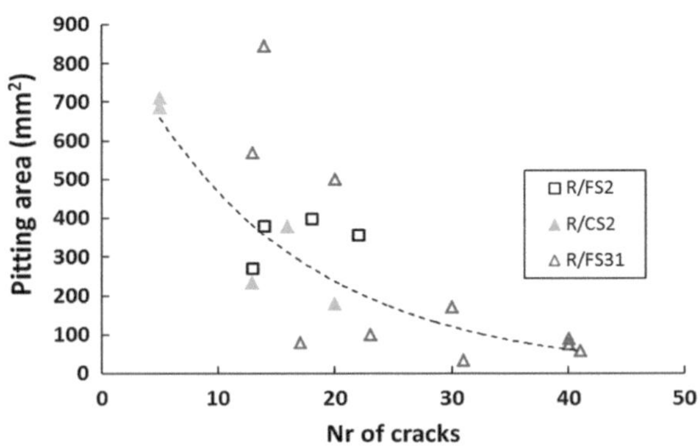

Fig. 9.16 Pitting area in steel, due to the different number of cracks in the R/SHCC specimens (Paul 2015)

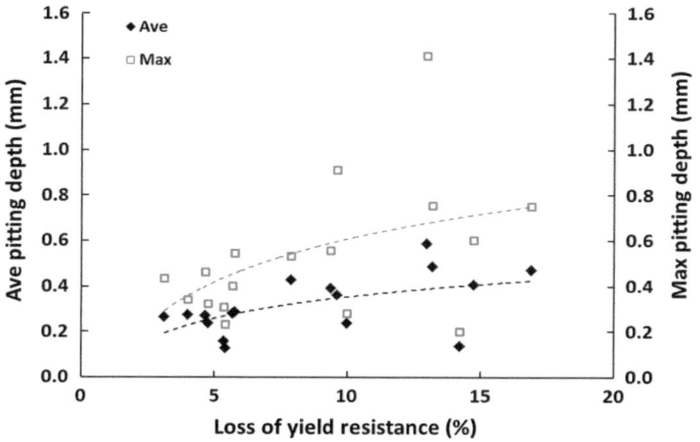

Fig. 9.17 Measured pitting depth versus the rebar loss of yield resistance (Paul 2015)

microscope, at the location of the known crack widths. A trend of decreased pitting diameters was found in the steel bars for relatively large total crack widths. Therefore, in terms of the corrosion damage of steel bars, the number of cracks, of total crack widths, and of crack spacing can be seen to be related.

Figure 9.17 shows a relationship between pitting depths and corrosion-induced loss of tensile yield resistance. An upward trend was found in the loss of yield resistance with increased pitting depth. Of importance in rebar resistance is the reduction in its cross-sectional area, which is clearly not represented by means of the pitting depth only, in the light of the rather poor correlations shown in Fig. 9.17. More accurate representation of the loss of the cross-sectional area is required to improve the correlation and estimation of corrosion damage to steel reinforcement, which is the subject of further research.

9.3.4 Crack Indices and Corrosion Damage

In Sect. 2.2, the IPI_{Cl} and *CWV* indices that were derived were subsequently applied to actual observed and experimentally obtained crack patterns and chloride profiles in Sect. 2.5.3. A reasonable correlation between the indices and the chloride content at the steel bar depth in cracked R/SHCC specimens exposed to a limited number of chloride wetting and drying cycles was shown in Fig. 2.21. However, the actual corrosion, in terms of average or maximum pitting corrosion depths, is not correlated with the indices, as is shown in Fig. 9.18 for maximum pitting depths. The above finding is to have been expected, as pitting corrosion was found to be uncorrelated to average or maximum crack widths in chloride-exposed specimens by Paul (2015), whereas the *CWV* is strongly correlated with average crack width, and the CWV_{WP} with maximum crack width, as is shown in Fig. 9.19.

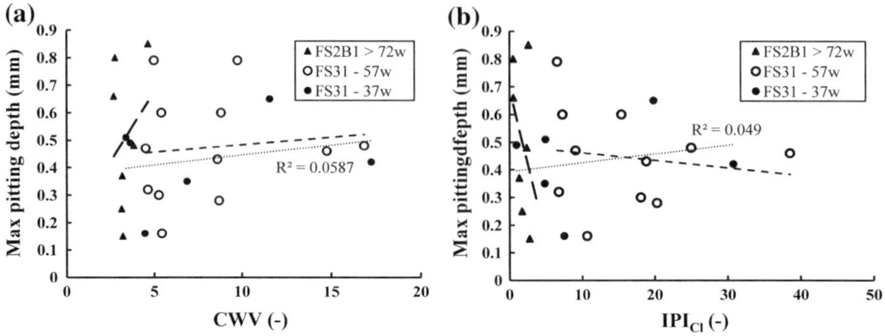

Fig. 9.18 Maximum pitting depth observed in corroded bars by Paul (2015), in Series 1 and 2 specimens (Table 2.2), showing no correlation with **a** *CWV* index, and with **b** IPI_{Cl}

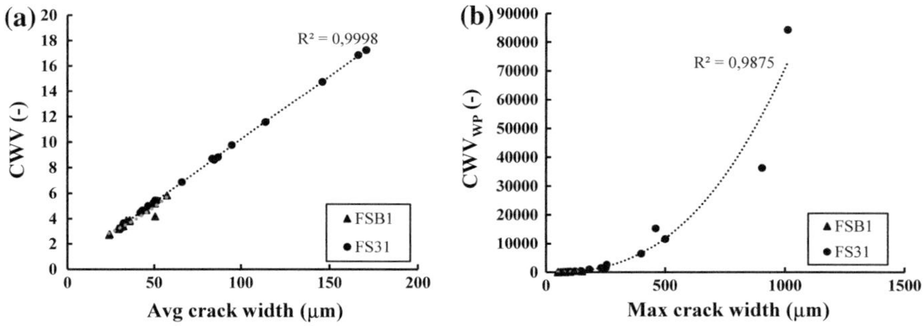

Fig. 9.19 Correlation between **a** the *CWV* and the average crack width, and between **b** the CWV_{WP} and the maximum crack width, for Series 1 and 2 in Table 2.2 (Paul 2015)

9.3.5 Splitting Corrosion Cracking

Grubb et al. (2007) performed chloride-induced corrosion experiments on lollypop specimens, which, in the present case, were 75-mm-diameter cylindrical specimens of a length of 150 mm, with a concentrically embedded steel bar of a diameter of 9.52 mm. Two water-to-cement ratio cement-based matrices (with a w/c of 0.4 and 0.55) were used. For each w/c ratio, a set of fibre-reinforced mortar specimens was prepared containing 4.5% by volume of steel microfibres. As control, a mortar mix was prepared for each w/c ratio. At the age of 28 days, all the specimens were exposed to a 3.5% sodium chloride, aerated aqueous solution, firstly for one day standing only up to 38 mm in the solution, after which the 150-mm-high lollypop specimens were fully submerged in the solution. Acid soluble chloride content tests, giving the total chloride content, were performed on half of the specimens after 22 weeks of submergence, showing that the fibre-reinforced specimens had higher total chloride, both in their surface regions and internally, than did the control mortar specimens for both w/c ratios. However, electrochemical testing during the 22-week submergence in NaCl solution resulted in significantly higher corrosion current densities (i_{corr}) for the control mortar specimens than for the fibre-reinforced ones. The authors postulate that the presence of microfibres close to the reinforcing bar served to control the splitting cracks arising from corrosion-product-induced swelling. Accordingly, the cracks concerned were eventually filled with corrosion products, which prevented further ingress of deleterious materials to the steel surface, keeping the corrosion rate low.

Jen and Ostertag (2016) performed chloride aqueous solution (3.5%) ponding on steel-reinforced beams of hybrid fibre-reinforced concrete (R/HyFRC), as well as on reinforced concrete (RC) control beams. The matrices of the control and HyFRC specimens contained the exact same amount of water, cement and fly ash content, with w/c = 0.6 and w/b = 0.45. In the HyFRC specimens, 0.2 vol% PVA fibres (L_f = 8 mm, d_f = 0.04 mm), together with 1.3 vol% hooked-end steel fibres (L_f = 30 mm, d_f = 0.55 mm) replaced some of the fine and coarse aggregate in the control specimen mixes. The cover-to-steel bar surface was 25 mm for both beam types. The small R/HyFRC and RC beams, with a cross-section of 152 × 152 mm, and 610 mm long, were subjected to the same flexural load, leading to one or two wide (0.2–0.3 mm) transverse cracks, and to a longitudinal splitting crack 130–250 mm long in the RC beams, but no visible cracks in the R/HyFRC beams. After roughly two years, during which continuous ponding was maintained in the first year, and while the solution was allowed to evaporate in the second year, electrochemical measurements showed the presence of significantly higher corrosion currents in the reference RC beams (i_{corr} = 2.80, and 6.05 $\mu A/cm^2$) than in the R/HyFRC beams (i_{corr} = 0.4, and 0.42 $\mu A/cm^2$). Visibly increased crack widths in the RC beams, as well as significantly higher steel mass loss in the reinforcing bars of 8 and 12 g respectively for the bars in the RC specimens, compared with 2 and 4 g respectively for the bars in the two R/HyFRC beams, confirmed the presence of higher corrosion rates in the RC control beams. As postulated by Grubb et al. (2007), Jen and Ostertag (2016) suggest that corrosion splitting crack control by the fibres involved led to the reduced corrosion rates concerned.

In the R/HyFRC beam specimens of Jen and Ostertag (2016), the corrosion of steel fibres was noted in the part of the sample that was directly adjacent to the pond of NaCl aqueous solution. Mihashi et al. (2011) also observed dispersed corrosion stains on their hybrid fibre-reinforced (V_f = 1.5% steel, together with 1.5% PE fibres) HFRCC beam specimens, in which they found significantly lower steel bar corrosion than they did in the R/mortar and R/FRCC (strain-softening) specimens containing only 1.5% PE fibres (see Fig. 9.11a). They postulate that the low corrosion rate observed in the steel-reinforcing bars embedded in HFRCC was caused by two mechanisms, namely crack width control by the fibre bridging, as well as the mechanism of steel fibres acting as sacrificial anodes.

9.4 Towards Corrosion Modelling in R/SHCC

Long-term experimental and field data are required for the development of reliable models. Until now, the data on SHCC has been limited in both field and laboratory experimental testing. As yet, no corrosion model exists for the new type of material concerned, with the existing corrosion models for NC also being criticised by researchers, as designers sometimes either overestimate, or underestimate, the amount of corrosion occurring in steel bars (Otieno et al. 2010). The highly complex nature of corrosion has resulted in there being no direct way of determining the actual corrosion rate. As a result, there are always some differences between the experimental findings and the actual corrosion that occurs in a structure. The present section describes a corrosion model developed by Paul (2015) for exploring single and multiple cracks in R/SHCC specimens.

9.4.1 Empirical Corrosion Modelling of R/SHCC

A corrosion model was proposed from the results that were obtained in for R/SHCC specimens. From the corrosion testing of R/SHCC, different correlations were observed between the corrosion, the cracks, the cover depth, and the chloride content. Based on the correlations, empirical Eqs. 9.2 and 9.3 were developed to predict the corrosion depths in the steel bars, due to the chloride-induced corrosion that was found to have occurred in the R/SHCC. Whereas Eq. 9.2 is valid for a single crack, Eq. 9.3 is valid for multiple cracks in the R/SHCC specimen. In addition, the corrosion rate that was measured by Paul (2015) was based on application of the coulostatic method, so that the equations in question may require adjustment before use for another corrosion rate measurement method. Figure 9.20 shows the experimental and model data for single-crack specimens with 15 and 25 mm cover depths. The multiple cracks, and the influence of crack spacing, that were also considered in the modelling of R/SHCC specimens are shown in

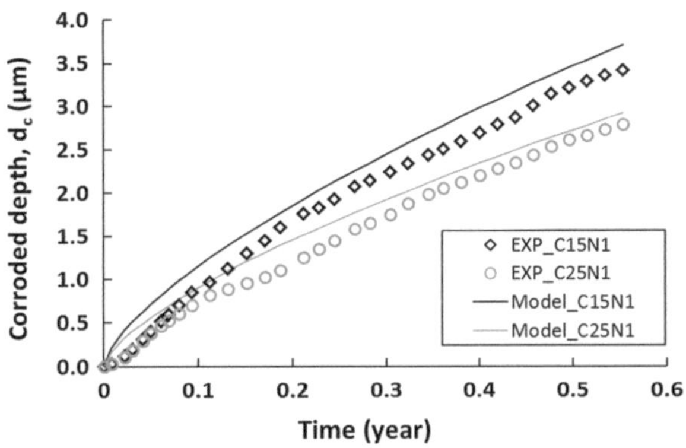

Fig. 9.20 Corrosion modelling using Eq. 9.2 of cracked R/SHCC specimens with a single crack (Paul and Van Zijl 2016)

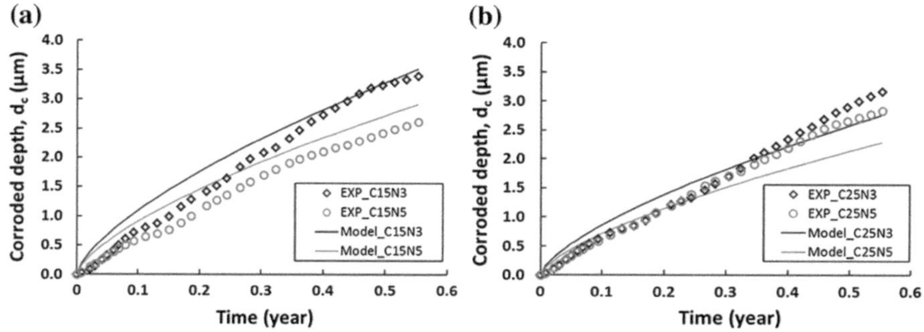

Fig. 9.21 Corrosion modelling using Eq. 9.3 of cracked R/SHCC specimens with multiple cracks (Paul and Van Zijl 2016)

Fig. 9.21a, b. As noted, the corrosion depth models are based on the experimental corrosion rate that was obtained from the Coulostatic method, and on the estimated corrosion depths that were calculated by means of integrating the corrosion rate over time (as in Eq. 9.1). However, there are several ways of determining the corrosion rate that occurs in a steel bar. Therefore, for different corrosion methods, the models may require calibration. Also, the influence of temperature, relative humidity (RH), amount of oxygen, and moisture content in the R/SHCC specimen were not considered in the modelling. Therefore, when the aforementioned parameters are known in the SHCC, the current modelling also requires improvement.

For single-crack specimens:

$$d_c = \frac{\delta_T \delta_{RH} W_{cr,ave} t^{\alpha_t} Cl_{free}^{\alpha_{cl}}}{C^{\alpha_C}} \tag{9.2}$$

For multiple-crack specimens:

$$d_c = \frac{0.35 \delta_T \delta_{RH} W_{cr,ave} t^{\alpha_t} Cl_{free}^{\alpha_{cl}} S_{cr}^{\alpha_S}}{C^{\alpha_C}} \tag{9.3}$$

with:

d_c the corroded depth (μm)
t the corrosion time (year)
$W_{cr,ave}$ the average crack width in the specimen (μm)
Cl_{free} the percentage of free chloride at the steel surface level (i.e. the % wt of the binder)
S_{cr} the average crack spacing (in mm) in the specimen
C the cover depth of the steel bar

The factors that depend on the temperature and on the relative humidity (RH) at the steel-bar level are δ_T and δ_{RH}. At the specimen surface temperature of 21 ± 2 °C, the δ_T can be considered to be 1. Similarly, for 55 ± 5% RH, δ_{RH} can be considered to be 1. The factors that depend on the ratio of wetting and drying periods, on the types of binder, on the range of maximum crack spacing, and on the quality of cover concrete (i.e. the microstructure and the tensile strength of concrete) are α_t, α_{cl}, α_S and α_C. In the current research work, the values of α_t, α_{cl}, α_S and α_C were considered to be 0.68, 0.5, 0.27, and 0.3, respectively. In terms of the values concerned, Fig. 9.20 shows reasonable agreement between Eq. 9.2 and the experimental data for single-crack (notched) FS32 specimens, whereas in Fig. 9.21a, b, Eq. 9.3 is shown to be in reasonable agreement with the corroded depths in the multiply cracked R/FS32 specimens. However, the values may need to be recalibrated for other material types, exposure conditions, ages, and different corrosion rate measurement techniques. In both Figs. 9.20 and 9.21, the data points represent Coulostatic corrosion rate measurements and the application of Eq. 9.1.

9.4.2 Durability Design Approaches for Chloride-Induced Corrosion in R/SHCC

At the current state of research into corrosion, the report on the corrosion rate of steel bars inside concrete is incomplete, and, in some cases, it is also contradictory. The pool of research results is limited, and no validation has been done by way of the actual field corrosion of an R/SHCC structure. In the present instance, experimental observations of the chloride-induced steel bar corrosion rates and the damage in R/SHCC are used as a basis for guidelines of the responsibilities of both the researcher and the durability designer:

- For the particular R/SHCC, the parameter values involved in the corrosion process require determining, including: (i) the crack width distribution; (ii) the crack spacing; (iii) the cover depth that may influence the crack width and spacing and the time taken for the oxygen, the free chloride, and the water to reach the steel-initiating corrosion and to enhance the corrosion rate; and (iv) the link between crack patterns and free chloride content at the level of the steel bar in the expected chloride exposure conditions.
- In laboratory experiments, and in the conditioned assessment of existing structures or structural elements, the corrosion rate requires estimating by means of such measuring techniques as the polarisation method, the Galvano static method, or the Coulostatic method.
- To improve the reliability of durability design in chloride exposure conditions, an increased data pool is required of non-destructive corrosion rate measurements, correlated with actual corrosion damage, such as pitting depth/area, mass loss, localisation of corrosion damage as a function of crack pattern, and loss in rebar resistance.

- The limit states of such parameters and the amount of damage that can be incurred before R/SHCC structures need repair or demolition require determination.

At the time of publication, no data on carbonation-induced corrosion in R/SHCC is yet available, which should be considered a worthy subject for further research.

9.5 Conclusions

This chapter has discussed the chloride-induced corrosion of reinforcement in SHCC, in particular when cracking occurs in the SHCC. Carbonation-induced corrosion in R/SHCC remains a research need.

In corrosion testing, greater estimated corrosion depth and actual measured and pitting depth in the steel were found in the specimens with a cover depth of 15 mm. A smaller difference was observed between specimens with 25 and 35 mm cover depths, suggesting a threshold cover thickness for specimens of this matrix type.

The average corrosion pitting depths and pitting areas in the steel bars are generally lower for R/SHCC specimens with smaller crack spacing. A reasonable correlation was found between the average pitting depths and the loss of yield force capacity of the steel bars, although localisation and cross-section reduction warrant further investigation.

The amount of free chloride content at the level of the steel bar surface appears to be better correlated to the actual corrosion damage than is the total chloride content. The experimental results of chloride-induced corrosion reported in this chapter can be represented by power functions incorporating average crack width, crack spacing, free chloride content at the steel surface, and time of exposure. Crack pattern indices developed in Chap. 2 appear to be uncorrelated to pitting depth, despite their reasonable correlation with chloride content at shorter exposure duration found in Chap. 2. It appears that crack spacing may dominate corrosion rate in cracked R/SHCC, due to the slow corrosion in the microcell chemo-electrical corrosion process.

References

Boshoff, W.P., Altmann, F., Adendorff, C.J., Mechtcherine, V. (2016). A new approach for modelling the ingress of deleterious materials in cracked strain hardening cement-based composites. Materials and Structures 49(6):2285-2295.

Broomfield J.P. (2007). Corrosion of steel in concrete understanding, investigating and repair. 2nd edition, Taylor & Francis, Milton Park.

Grubb, J.A., Blunt, J., Ostertag, C.P., Devine, T. (2007). Effect of steel microfiber on corrosion of steel reinforcing bars. Cement and Concrete Research 37(7):1115-1126.

Jen, G., Ostertag, C.P. (2016). Experimental observations of self-consolidated hybrid fiber reinforced concrete (SC-HyFRC) on corrosion damage reduction. Construction and Building Materials 105:262-268.

Kobayashi, K., Iizuka, T., Kurachi, H., Rokugo, K. (2010). Corrosion protection performance of high performance fibre reinforced cement composites as a repair material. Cement and Concrete Composites 32:411-420.

Maalej, M., Ahmed, S.F.U., Paramasivam, P. (2003). Corrosion durability and structural response of functionally-graded concrete beams. Journal of Advanced Concrete Technology 1 (3):307-316.

Mihashi, H., Ahmed, S.F.U., Kobayakawa, A. (2011). Corrosion of reinforcing steel in fibre reinforced cementitious composites. Journal of Advantage Concrete Technology 9(2):159-167.

Miyazato, S., Hiraishi, Y. (2005). Transport properties and steel corrosion in ductile fibre reinforced composites. Proceedings of 11th International Conference on Fracture (ICF11), 20-25 March 2005, Torino, Italy, Politecnico di Torino, Torino, CD-ROM.

Miyazato, S., Hiraishi, Y. (2013). Durability against steel corrosion of HPFRCC with bending cracks. Journal of Advantage Concrete Technology 11:135-143.

Mohammed, T.U., Otsuki, N., Hisada, M., Shibata, T. (2001). Effect of crack width and bar types on corrosion of steel in concrete. Journal of Materials in Civil Engineering 13:194-201.

Otieno, M.B., Alexander, M.G., Beushausen, H.D. (2010). Corrosion in cracked and un-cracked concrete influence of crack width, concrete quality and crack reopening. Magazine of Concrete Research 62(6):393-404.

Paul, S.C., Van Zijl, G.P.A.G. (2014). Crack formation and chloride induced corrosion in reinforced strain hardening cement-based composite (R/SHCC). Journal of Advanced Concrete Technology 12:340-351.

Paul, S.C. (2015). The role of cracks and chloride in corrosion of reinforced strain hardening cement-based composite (R/SHCC). PhD thesis, Stellenbosch University, South Africa.

Paul, S.C., Van Zijl, G.P.A.G. (2016). Chloride-induced corrosion modelling of cracked reinforced SHCC. Archives of Civil and Mechanical Engineering 16:734-742.

Şahmaran, M, Li, M., Li, V.C. (2007). Transport properties of engineered cementitious composite under chloride exposure. ACI Materials Journal 104(6):243-250.

Şahmaran, M., Li, V.C., Andrade, C. (2008). Corrosion resistance performance of steel reinforced engineered cementitious composite beams. ACI Materials Journal 105(3):604-611.

Song, G., Shayan, A. (1998). Corrosion of steel in concrete: causes, detection and prediction. A state-of-art-review, ARRB Transport Research Ltd, Review Report 4.

Wagner, C. (2016). Dauerhaftigkeitsrelevante Eigenschaften von dehnungsverfestigenden zementgebundenen Reparaturschichten auf gerissenen Betonuntergründen (Durability-related properties of strain-hardening cement-based repair layers on cracked concrete substrates). Doctoral thesis, Technische Universität Dresden, Germany.

Chapter 10
Durability and Service Life Design Concepts for Structures and (Non-)Structural Members Made of or Strengthened/Repaired with SHCC

Viktor Mechtcherine, Frank Altmann and Gideon P.A.G. van Zijl

Abstract Due to their unique properties, strain-hardening cement-based composites (SHCC) are well-suited for structural and non-structural members, as well as for the repair and rehabilitation of ordinary concrete structures that are exposed to severe mechanical or environmental loading. To fully utilise their advantageous durability properties, a comprehensive durability and service life design framework is required. This framework must encompass design concepts for all relevant load cases and load case combinations. For an efficient service life design of members and structures, the design concepts must be performance-based. In this chapter, SHCC applications and relevant degradation processes are classified. Following the classification, possible service life design approaches are described, categorised and assessed regarding their suitability for the service life design of SHCC members and structures. Following an outline of a possible development strategy for a service life design framework, available design concepts for chloride-induced rebar corrosion are presented. It is shown that in order to develop available concepts into a comprehensive durability design framework, the uncertainty of input variables must be reduced, and deterioration models must be further validated. Finally, other limit states, load cases, and load case combinations must be considered. Moreover, it is postulated that for most load cases and load case combinations, suitable deterioration models for SHCC with reasonably well-quantified input variables will require significant additional development. Hence, deterioration model-free durability assessments with fuzzy uncertainty quantifications will likely be the most suitable option for most load cases and load case combinations for the foreseeable future.

Keywords Durability design · Service life design · Performance-based design · Uncertainty · Probability theory · Fuzzy logic · Fuzzy randomness · Subjective information

---

V. Mechtcherine (✉) · F. Altmann
Technische Universität Dresden, Dresden, Germany
e-mail: viktor.mechtcherine@tu-dresden.de

G.P.A.G. van Zijl
Stellenbosch University, Stellenbosch, South Africa

G.P.A.G. van Zijl and V. Slowik (eds.), *A Framework for Durability Design with Strain-Hardening Cement-Based Composites (SHCC)*, RILEM State-of-the-Art Reports 22,
DOI 10.1007/978-94-024-1013-6_10

10.1 Durability Requirements

Strain-hardening cement-based composites (SHCC) are characterised by unique mechanical and durability properties that not only give them the potential to improve member durability, but increasingly allow new structural solutions not feasible with ordinary concrete (Lepech and Li 2009; Mechtcherine 2013). Thus it is to be expected that their application will be targeted at structural and non-structural members, as well as at the repair and rehabilitation of ordinary concrete structures that are exposed to severe mechanical or environmental loading (Li 2003; Mechtcherine and Altmann 2011).

Applications may be classified as suggested in Table 10.1. For each individual application, the specific durability requirements will depend on the individual structure or member and its exposure conditions.

For any structure and member to exhibit the required durability, the durability of the applied materials must first be assured. As indicated in Table 10.1, structural SHCC members and reinforced concrete (RC) elements with SHCC repair layers are likely to contain steel reinforcement bars (Mechtcherine 2012; Mechtcherine and Altmann 2011).

Hence, the following durability requirements arise:

i. the protection of steel reinforcement from corrosion;
ii. the durability of the SHCC matrix;

Table 10.1 Classification of SHCC applications exposed to environmental actions (Altmann 2012)

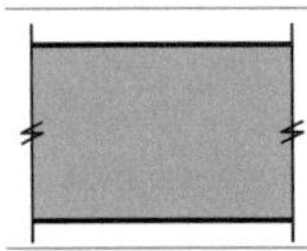	Currently, no field applications of unreinforced SHCC members exist, but non-structural members may well contain no additional bar reinforcement.
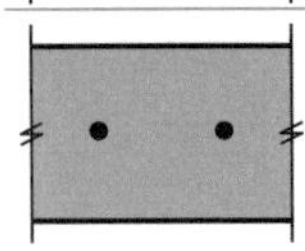	SHCC members with (steel) reinforcement bars (R/SHCC) such as link slabs (Lepech and Li 2009; Mechtcherine 2013) may provide alternatives to installed parts (e.g. in bridge joints), or act as plastic hinges under earthquake loading, thus providing more robust structures and superior post-disaster load capacities for critical infrastructure such as road bridges or harbour structures.
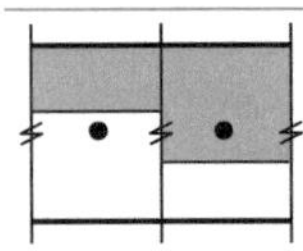	(Un)reinforced SHCC overlays for ordinary concrete are already being used in Japan for the waterproofing of dams and irrigation channels. The material has also been successfully used as an overlay of cracked RC structures such as retaining walls, railway viaducts, and tunnel linings (Rokugo and Kanda 2013; Rokugo et al. 2009; Yamamoto et al. 2005). Furthermore, SHCC overlays have been used for the patch repair of concrete road surfaces (Wagner et al. 2008).
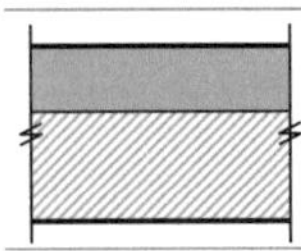	Unreinforced SHCC overlays over other material are also possible. Realised applications include the waterproofing of non-concrete irrigation channels and the steel–SHCC composite bridge deck of Mihara bridge, both in Japan (Kunieda and Rokugo 2006; Rokugo et al. 2009).

iii. fibre durability; and
iv. the durability of fibre–matrix bond properties.

Key degradation mechanisms associated with the above requirements are outlined in Fig. 10.1.

Since SHCC is characterised by strain-hardening tensile behaviour due to the formation of multiple cracks of limited width under tensile load, the material is expected to be used in members that will typically become cracked under service loads. The limited crack width of SHCC suggests that rebar corrosion is initiated later and that deterioration in the form of steel cross-section reduction, cracking, and spalling is slower in R/SHCC members than in ordinary RC members exposed to the same mechanical and environmental loads.

The durability of fibre, matrix and fibre–matrix bond properties is absolutely essential with respect to ensuring the key performance property (i.e. the ductility) of SHCC. Since this property is achieved by means of a comprehensive material design that builds upon a balance of properties of the individual components (fibre, matrix and interface). Any disproportionate change in the properties of these components — resulting from any kind of environmental loading — may lead to the worsening of the tensile behaviour of the SHCC, which can be considered as deterioration. Obviously, the loss of ductility also means a change in the crack pattern of SHCC, in other words fewer but wider cracks. This compromises the beneficial effect that can be attained by strongly limiting crack widths, which is, in fact, the main argument for the use of SHCC with respect to the durability of structural members. The transport properties and thus the resistance of SHCC in aggressive environments are strongly affected by its cracking width and spacing characteristics, as was discussed in Chap. 2. The same holds true, irrespective of whether an SHCC member contains steel reinforcement or not.

While the requirements for the durability of the fibre and the interface between the fibre and cement-based matrix are limited to ensuring the ductility of SHCC throughout the service life of a SHCC element, the durability of the matrix is an additional requirement to ensure structural integrity, bond with reinforcement, aesthetics, and so forth. Here, incidences of exposure and the general requirements imposed on the matrix are the same as for ordinary concrete. However, the performance can differ significantly, as has been outlined in previous chapters.

In addition to these material-centric considerations, degradation processes at the member or structural level may require consideration. These include the time-dependent bonding of overlay systems (cf. Chap. 8), as well as the durability of joints between SHCC and non-SHCC elements (cf. application examples in Chap. 1).

A comprehensive review and discussion of the durability of SHCC, in general, and structural elements and structures, in particular, can be found in Mechtcherine (2012), Mechtcherine and Altmann (2011), Van Zijl and Wittmann (2011) and Van Zijl et al. (2012).

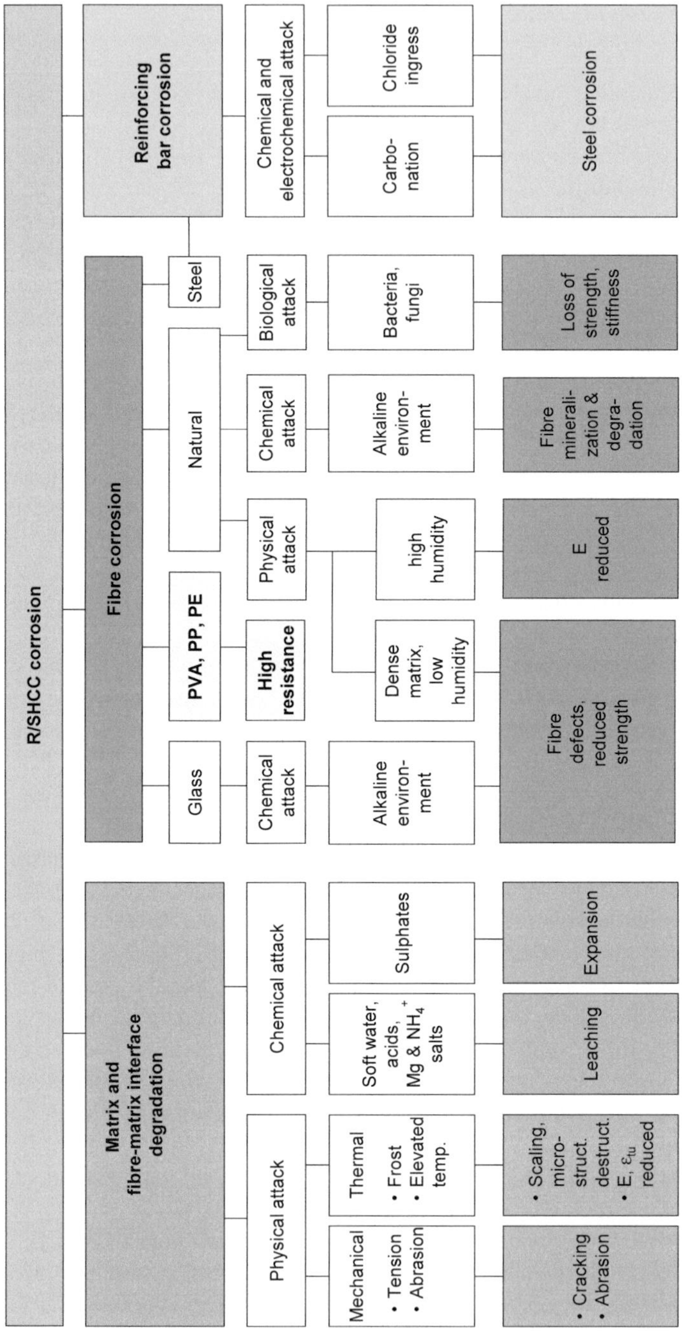

Fig. 10.1 Overview of degradation processes for steel-reinforced SHCC (R/SHCC) members

10.2 Durability and Service Life Design Approaches

To fully utilise the durability properties of SHCC for the efficient design of members and structures, a comprehensive durability design concept is required. The concept concerned must be appropriate for the envisaged applications, in the light of the available information on the relevant deterioration processes involved.

Durability is defined as "the capability of a structure to maintain minimum performance under the influence of (mechanical and environmental) loads" (Söderqvist and Vesikari 2004), and durability design, or service life design, ensures that the performance in question is achieved over the intended period of time (i.e. its service life). Since any structure consists of several members, or elements, the consequences of the deterioration of one or more of its components, or members, for a structure's durability must be assessed (e.g. by means of using fault tree analysis) (Müller and Vogel 2008) (cf. Fig. 10.2). At the core of such analysis are design concepts on the member level. However, it must be stressed that service life design begins with sound architectural and structural design and detailing, including the choice of appropriate construction materials (Altmann 2010).

On the member level, two fundamentally different design strategies must be differentiated from each other (cf. Fig. 10.3). On the one hand, an environmental load may be avoided, for instance, by means of coating the member, or the component, with an impervious barrier, or by means of replacing a component, such as mild steel with stainless steel. On the other hand, the steel-reinforced SHCC (R/SHCC) member may be designed to withstand the environmental load. In the latter case, Gehlen (2000) differentiates between three different design levels. On the micro level, which is unsuitable for engineering applications, the deterioration processes are modelled scientifically correctly. On the meso-level, which is appropriate for monumental structures with a service life of 100 years or more, empirical models are used to model the deterioration concerned. Finally, on the macro-level, simple deemed-to-satisfy concepts are applied to ensure the required service life.

Most current design codes, like Eurocode 2 (DIN EN 1992-1-1:2011-01), in combination with relevant material codes as DIN EN 206-1:2001-07, prescribe a deterministic design approach. Environmental conditions are classified into a limited number of exposure classes, with deemed-to-satisfy rules being prescribed for

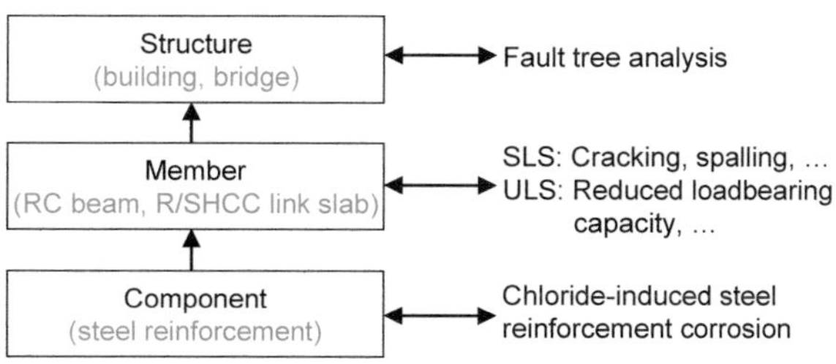

Fig. 10.2 Consequences of steel reinforcement corrosion

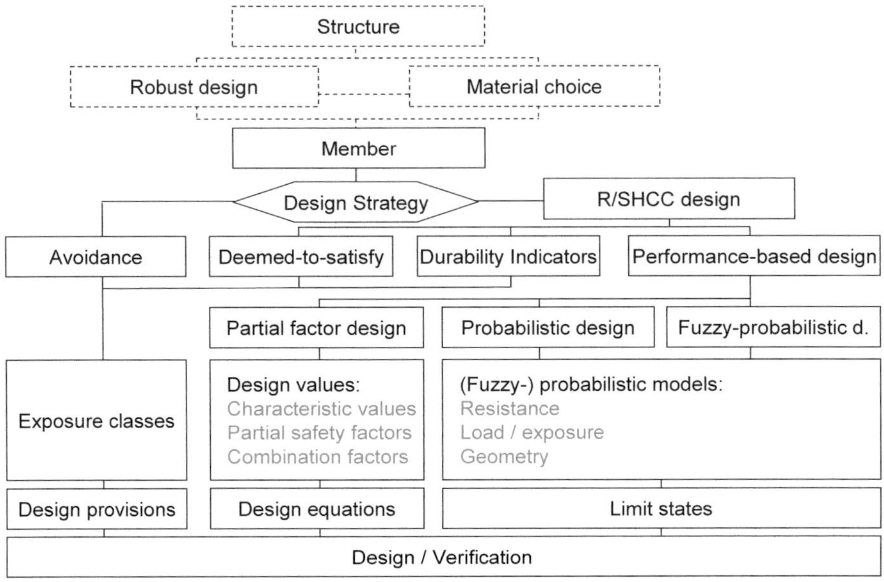

Fig. 10.3 Different levels of sophistication for service life design, as adapted from Altmann (2010) and from the Model Code for Service Life Design (fib 2006)

each class. Structures that are designed according to such rules are likely to have an acceptably long, though unspecified, lifetime (DAfStb 2008), even though a service life of 50 years is implicitly assumed within the Eurocode 2 framework (DIN EN 1990:2010-12; DIN EN 1992-1-1:2011-01).

Durability design has long been recognised as using empirical requirements that are based not on scientific principles, but on field experience that is combined with limited research data and with expert judgement, which amounts to an unsatisfactory situation (Edvardsen 2010). Such practice has given rise to the development of relatively sophisticated design concepts.

A first step was taken with the development of performance, or durability, indicators (Alexander et al. 2008; Baroghel-Bouny et al. 2011). In contrast to the deemed-to-satisfy rules, durability indicators are based on established relations between characteristic laboratory-measurable properties and the deterioration process in question. Indicators include, for instance, mass loss due to scaling after frost or freeze–thaw exposure, and chloride diffusion, or migration, coefficients. To avoid inconsistencies, Gehlen and Von Greve-Dierfeld (2010) suggest differentiation between empirical indicators, which usually assume a service life of 50 years, and indicators that are quantified on the basis of probabilistic analyses.

The development of sophisticated design concepts culminated in performance-based design concepts with probabilistic and fuzzy-probabilistic input variable quantifications, as is discussed in the following section. In light of the limitations of the deemed-to-satisfy approach, some recent design or materials standards such as DIN EN 206-1:2001-07 are open to the application of such performance-based design concepts, especially for the design of structures with a service life of 100 years and longer. Using a limit-state approach, the above-mentioned concepts allow for a systematic member design to ensure that, under the actual environmental loads experienced, the specified service life is reached with the desired reliability.

10.3 Performance-Based Design

10.3.1 Fundamentals

According to Gehlen (2000), Gehlen et al. (2008) and Altmann (2012), a performance-based durability design concept requires:

i. a deterioration model, e.g. a constitutive equation describing the deterioration process in question realistically, and with an appropriate degree of sophistication;
ii. a reliable quantification of the deterioration model's input variables for typical ranges of mixture compositions, utilising an appropriate uncertainty model to adequately reflect the uncertainties associated with the model and the variable quantifications; and
iii. the formulation of appropriate limit states and the required reliability with which these are not exceeded.

The challenge that is posed by the above-mentioned requirements is highlighted by the fact that, despite decades of research and field application of ordinary RC, they have, as of yet, only been met for chloride- and carbonation-induced steel reinforcement corrosion.

More detailed information than is made available here on the principles of probabilistic design, on its application to service life design, and, more specifically, on the DuraCrete framework and the concepts derived there, can be found in: DAfStb (2008), DuraCrete (2000a, b, c), fib (2006, 2010), Gehlen (2000), Gehlen and Kapteina (2004), Gehlen et al. (2008), ISO (2012), JCSS (2001a, b, c), and Lay and Schießl (2003).

10.3.2 Probabilistic Design — Limit States and Reliability

In performance-based design, loads or stresses S and resistances R are juxtaposed, and failure is assumed to occur if the load exceeds the resistance involved. In this case, the limit state function that is given in Eq. 10.1

$$Z = R - S \tag{10.1}$$

assumes a negative value.

Since neither load, nor resistance are ever fully known, and since they both exhibit considerable scatter, the associated uncertainty must be considered in the design process. To this end, the input variables S and R must be quantified, using an appropriate uncertainty model (cf. Sect. 10.4). One option is a probabilistic variable quantification. In such a case, the probability of failure, as defined above, can be computed, according to Eq. 10.2, as

$$p_f = p\{Z = R - S \leq 0\} \leq p_{f,\max} \tag{10.2}$$

For the durability to be sufficiently high, the failure probability must be smaller than an acceptable probability $p_{f,\max}$, over the intended service lifetime of the member. Alternatively, a minimum value of the reliability index β (cf. Fig. 10.4) may be specified.

Both load (S) and resistance (R) are generally functions of space and time. Typically, the uncertainty of predictions of both variables increases over time, which results in the flattening and widening of probability density functions $f(s)$ and $f(r)$. Also, the mean values of both variables can change (Melchers 1987). Hence, the safety index β typically decreases over the service life of a structure.

Further information on the theory and mathematics of reliability theory can be found in: Lay and Schießl (2003), Lee and Hwang (2008), Melchers (1987), Schuëller (1997), and Schuëller et al. (2003).

The stochastic quantification of the relevant input variables requires a very large database. Due to the large number of influencing factors, the key variables of the constitutive equations must be quantified separately for different influencing parameters including, but not limited to: (i) binder type; (ii) binder content; (iii) water-to-binder ratio; and (iv) exposure conditions. For (R/)SHCC, such additional parameters as (v) the fibre type, and (vi) the crack distribution (cf. Chap. 1) must be considered for separate variable quantification. The latter may,

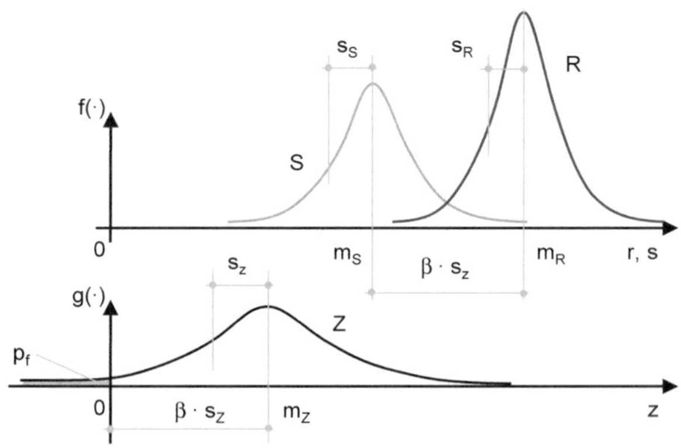

Fig. 10.4 Normally distributed load S, resistance R, and reliability Z with the means m and standard deviations s, with the failure probability p_f and reliability index β also shown

however, be considered a material property that is realised upon exposure to an external load, and which, as such, can be a function of the parameters (i) to (v). In this case, parameter (iv) must reflect both future exposure (i.e. the traditional definition of exposure conditions), as well as previous exposure (i.e. the exposure history). Where a separate quantification for different influencing parameters is not possible (e.g. due to a lack of available information, or where parameters are quantified based on a mix of laboratory and field data), the scatter involved tends to increase significantly (Gulikers 2007).

10.3.3 Semi-Probabilistic Design

A probabilistic service life design requires sophisticated software and is too cumbersome for ordinary structures. However, the probabilistic concept can be transformed into a semi-probabilistic partial factor concept. The limit state function

$$R_d - S_d \geq 0 \tag{10.3}$$

uses deterministic design values denoted by the index *d*, which are computed as

$$R_d = \frac{R_c}{\gamma_R} \tag{10.4}$$

$$S_d = \gamma_S \cdot S_c \tag{10.5}$$

from characteristic values that are denoted by the index *c*, and from partial safety factors γ. To this end, the probabilistic quantification of input parameters must be translated into deterministic characteristic values and safety factors. Such translation requires the specification of a target reliability β, and of the design situations, for which the variables must be valid, as well as the information about the importance of the individual variables regarding the failure probability (Lay and Schießl 2003). The determination and validation of such variables is described by Gehlen and Von Greve-Dierfeld (2010).

For ordinary RC, the DuraCrete concept (DuraCrete 2000c) includes a semi-probabilistic approach that has been further developed into the Dutch CUR Guidelines (2009). However, doubts exist about the quantification of input variables, since different combinations yield very different reliabilities, which are not evident to the user (Gulikers 2011). Since, for any new material, significantly less information is available than there is for ordinary RC, it will be, at best, challenging to define suitable limit state functions and to reliably quantify input variables.

10.3.4 Limitations

Further to the limitations of the constitutive equations and the input variable quantification discussed above, criticism has been levelled at lack of explicit inclusion of the general workmanship onsite (labcrete vs. realcrete) and of the skin effect (covercrete vs. corecrete) in the constitutive equations. However, such factors are typically pragmatically included through the appropriate quantification of input variables.

More importantly, the constitutive equations neither allow for the consideration of load combinations that may well be critical, nor do they account for the 'size effect', according to which — as is the case with relatively long chains — comparatively large members tend to have a higher failure probability than do the other members (DuraCrete 2000b).

10.4 Uncertainty and Models Describing Uncertainty

10.4.1 Definition and Classification of Uncertainty

In the previous section, uncertainty and its implication for the service life have been discussed without elaboration as to the nature of this phenomenon. Firstly, a distinction should be made between data uncertainty and model uncertainty. Models, as approximations of reality, do not describe the actual processes with absolute certainty. Such is especially true for empirical deterioration models that are mathematical fits to observations, and that lack clear physical or (electro)chemical meaning. The uncertainty that is associated with such approximation of reality is referred to as 'model uncertainty'. In contrast, the uncertainty that is associated with the input variables is referred to as 'data uncertainty' (Möller et al. 2003).

Secondly, due to the difference in their nature, it is useful to distinguish between aleatory and epistemic uncertainty. Aleatory, or stochastic, uncertainty, which is also referred to as 'randomness', is a result of the inherent, nondeterministic nature of a variable. Imprecision or epistemic uncertainty, in contrast, is the result of a lack of information, emanating from an insufficient number and/or imprecise measurements.

10.4.2 Probability Theory

Probability theory is the mathematical theory of chance. The likelihood, or probability $p(A)$, that an event A will occur is quantified by means of a single number between zero and one, with the largeness of the $p(A)$ indicating the likelihood of the occurrence of event A. If it is impossible that A occurs, its probability is $p(A) = 0$, whereas a certain event has a probability of one. Due to the bivalent nature of probability theory, the probability that the event A will happen $p(A)$ and the probability that its opposite will happen $p(not\text{-}A)$ always add up to 1 (Kosko 1993).

The uncertainty of a continuous variable x is quantified by means of the probability density function $f(x)$. The variables of the functions concerned are usually determined using estimation theory and an extensive database. If no a priori information is available, this may be done using the maximum-likelihood method, or by means of bootstrapping, whereas Bayesian methods are used when a priori information is available (Möller and Beer 2004).

10.4.3 Fuzzy Logic and Fuzzy Set Theory

The U.S. Department of Commerce (1991) defines fuzzy logic as follows:

> Fuzzy logic is a concept derived from the branch of mathematical theory of fuzzy sets. Unlike the basic Aristotelian theory (or bivalent logic) that recognises statements as only 'true' or 'false', or '1' or '0' (…), fuzzy logic is capable of expressing linguistic terms such as 'maybe false' or 'sort of true'. In general, fuzzy logic (…) allows (…) to emulate the human reasoning process, quantify imprecise information, make decisions based on vague and incomplete data, yet by applying a 'defuzzification' process, arrive at definite conclusions.

In fuzzy set theory, the degree of likelihood that a statement is true is quantified with a membership function μ, which may assume values between zero, if there is no possibility that a statement is true or an event occurs, and one in the case of maximum possibility. No further constraints exist for the shape of membership functions, which also allows for the quantification of such linguistic parameters as *very low*, *low*, *medium*, *high*, and *very high*. In contrast to probability theory, the sum of the possibility of an event and its complementary event does not need to be unity.

10.4.4 Theory of Fuzzy Randomness

Probability and fuzzy set theory may be combined to model the uncertainty that is associated with a variable x as being fuzzy random. In terms of such an approach, the parameters of the probability density function $f(x)$, such as mean value and standard deviation, are quantified as fuzzy parameters. Denoting fuzzy uncertainty with a tilde $\sim$, this yields fuzzy probability density functions $\tilde{f}(x)$ (cf. Fig. 10.5), and corresponding fuzzy probability distribution functions $\tilde{F}(x)$.

Further information on fuzzy randomness, its mathematical description, and the operation with fuzzy random variables and uncertainties can be found in: Buckley (2005), Möller and Beer (2004), Sickert (2005), and Viertl and Hareter (2006).

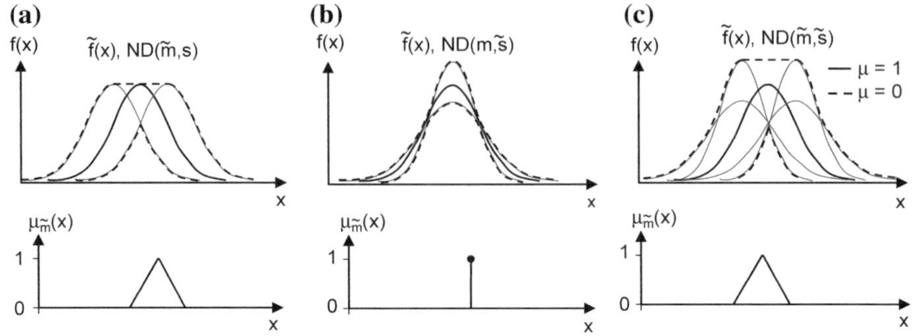

Fig. 10.5 Fuzzy probability density functions $\tilde{f}(x)$ and membership functions μ for its mean value x, for a normally distributed variable x with **a** a fuzzy mean value and a certain standard deviation, **b** a certain mean value and a fuzzy standard deviation, and **c** a fuzzy mean and a fuzzy standard deviation

10.5 Conceptual Framework for the Durability and Service Life Design of SHCC and R/SHCC Structures and Members

10.5.1 Preliminary Remarks

In the current section, options for a comprehensive framework for durability and service life design, as well as a strategy for its development, are outlined. Subsequently, available design concepts for chloride-induced rebar corrosion are discussed in more detail. Being the first available design concepts, they may provide valuable guidance towards the development of a comprehensive durability framework.

10.5.2 General Considerations

As outlined above, such a comprehensive framework is required for the efficient design of members and structures that fully utilise the beneficial properties of SHCC. The concepts in question, which are also useful in the context of materials research, may be categorised according to the degree of sophistication of their deterioration and uncertainty models.

According to Fig. 10.6, three general types of deterioration models can be identified:

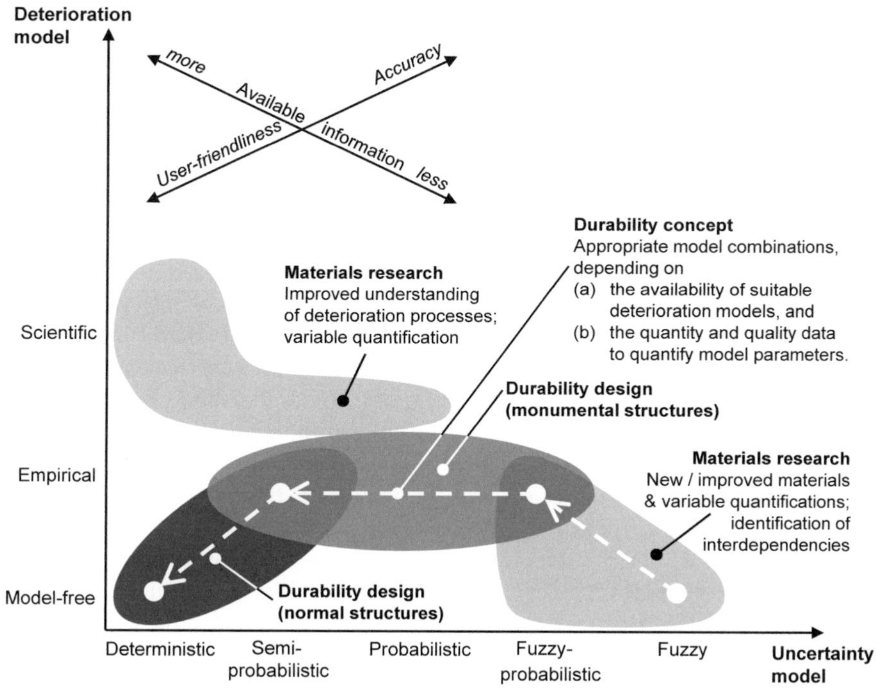

Fig. 10.6 Combinations of deterioration and uncertainty models, adapted from Altmann (2012), with the *dashed arrows* showing a reasonable path for the development of a durability concept over time

i. model-free approaches, as used in the deemed-to-satisfy design concepts that were discussed in Sect. 10.2;
ii. empirical models, consisting of typically simple analytical formulas that are mathematical fits to observations and that have no clear physical or (electro)chemical meaning; and
iii. scientific models describing the deterioration process in as much detail as it is currently known and understood.

The more sophisticated the model is, the more accurately it describes the actual physical and (electro)chemical deterioration processes involved. However, the price for attaining such accuracy is complexity and a lack of user-friendliness, which renders scientific models only suitable for research and for improving the level of confidence in empirical models, but not rendering them suitable for practical design applications (Nilsson 2005).

The design rules of deemed-to-satisfy concepts, in contrast, are simple and user-friendly. However, they cannot ensure that a desired service life is reached with a specified reliability, and even to ensure an acceptably long but unspecified service life, they require long-term experience not available for new materials.

Between the two extremes described, empirical deterioration models provide a good balance between accuracy and user-friendliness, and thus a rational basis for durability design.

The uncertainty of deterioration models and their input variables is disregarded if input variables are quantified deterministically. However, if the desired service life shall be reached with a specified reliability, performance-based limit state design concepts must be used, which require a quantification of this uncertainty. Probabilistic or semi-probabilistic uncertainty models require very large databases that include information on the long-term behaviour of the material involved. Such information is not yet available for SHCC, or for any other new material.

Such lack of information or non-stochastic uncertainty can be transparently modelled using fuzzy sets and fuzzy probability theory. In recent years, significant advances have been made in the development of methods and algorithms to analyse lifetime processes exhibiting both stochastic and non-stochastic uncertainties. Fuzzy stochastic analysis, for instance, allows the computation of results for empirical chloride ingress models with input variables exhibiting fuzzy stochastic uncertainty (Altmann 2012).

In the absence of deterioration models, neural networks may be used to predict fuzzy and fuzzy random time-dependent processes (Möller and Beer 2004; Reuter 2006; Sickert 2005). However, similar constraints as for scientific deterioration models apply to model-free approaches to quantify uncertainty.

In contrast, causal reasoning of the 'if–then' type, with a fuzzy quantification of the link between cause and effect, provides a transparent, yet deterioration model-free way of quantifying all available information (cf. Sect. 10.6). Typically, such an approach is likely to be limited to materials research. However, in the absence of any deterioration model, a better balance might be struck thereby between accuracy and user-friendliness than can be achieved by using the aforementioned neural network approach.

Further to the above, it must be noted, that current state-of-the-art durability or service life design concepts like the fib Model Code for Service Life Design (2006) for ordinary crack-free concrete or Altmann's (2012) design concept for SHCC only consider such individual load cases as chloride-induced rebar corrosion. The effects of other environmental, or mechanical, loads to which a member or structure is, or was, exposed are ignored, despite the fact that individual load cases may mutually reinforce their corrosive actions. Hence, such detrimental load combinations merit some attention.

10.5.3 Development Strategy

Following the example of the aforementioned existing design concepts, it seems reasonable to follow a modular step-by-step approach towards developing a comprehensive durability design framework for SHCC, as is outlined below, and which is described in more detail by Altmann (2012). As can be seen in Fig. 10.1, a significantly larger number of relevant degradation processes may be relevant for the service life of (R/)SHCC members compared with the number of such processes relevant for the service life of ordinary RC members. In the light of the limited amount of information that is currently available on the degradation of any new material and of the limited number of researchers investigating such matters, a significant challenge is posed.

In the light of the above, it is, thus, crucial that all available information is harnessed to the fullest extent, beginning with the identification of the most critical degradation processes involved. These processes should be given priority in the development of individual durability or service life design modules for each load case. If, and as, the required information becomes available, individual modules should be combined and expanded upon to account for load case combinations (Altmann 2012; Wittmann 2011).

As has been touched upon in previous sections and chapters, (R/)SHCC members are expected to be cracked under service loads. Hence, cracks, their time-dependent development (including their possible self-healing), their influence on the ingress of deleterious substances and on the propagation of deterioration processes, as well as the influence of deterioration processes on cracks will always require consideration.

Cracking is induced by one or more external loads. A priori, meaning in the design stage, the time-dependent crack pattern, hence, requires determination by means of a deterioration model. Thus, strictly speaking, each deterioration process, as identified in Fig. 10.1, consists of at least two load cases, namely the crack-inducing load case, and the deterioration-inducing load case. However, for the sake of simplicity, each deterioration process is referred to as an individual load case in the current chapter.

For each individual load case, there is, at least initially, no deterioration model, and there are typically only limited results of experimental investigations to enable the quantification of the degradation process and its time-dependent development. At the present stage, causal reasoning of the 'if–then' type is likely to be the only option that is available for describing the deterioration process. In Fig. 10.6, the option is represented by means of the white circle at the base of the rightmost arrow. Section 10.7 discusses some tools that are available for transparent causal reasoning that is based on subjective information.

As more information than is available at present comes to light, deterioration models for the overall degradation process, or aspects thereof, may be developed from scratch, or by means of adapting models that have been developed for such similar materials as ordinary concrete. While aspects of the deterioration process may be well enough understood to be modelled scientifically, it is likely that the initial models that are used to describe the overall degradation process will be empirical. Similarly, limit states may be adapted from other materials, or else newly defined, especially in the case of such new deterioration processes as those that are associated with natural fibres (cf. Fig. 10.1 and Chap. 3).

As is to be expected, currently little information is available about the deterioration process in question. This information deficit results in non-stochastic uncertainty atop irreducible stochastic uncertainty, with the combination of such uncertainties best being quantified by means of a fuzzy-probabilistic uncertainty model. The potential of such a design concept notwithstanding, its location in Fig. 10.6 at the tip of the rightmost arrow highlights its limited applicability to real-life structures, due to the elaborate uncertainty model. Thus, the database requires improvement as a step towards a user-friendly, semi-probabilistic design concept.

A targeted and efficient reduction of model and parameter uncertainty requires:

i. the identification of key input variables, with the help of sensitivity analyses;
ii. experimental investigation;
iii. investigations of field applications; and
iv. numerical studies (e.g. the use of neural networks that are focused on the identified variables).

With significant expansion of the pool of available information, the amount of both model and parameter uncertainty can be expected to shrink over time, allowing for a purely stochastic variable quantification. As has been shown, for instance, in the case of the DuraCrete (2000a, b, c) project, at this stage a semi-probabilistic parameter quantification becomes both possible and meaningful. Such development is represented by means of the middle arrow in Fig. 10.6.

In principle, it will also be possible to develop a deemed-to-satisfy design concept (cf. Fig. 10.6, tip of the leftmost arrow), once enough experience has been gained in terms of the probabilistic or semi-probabilistic concept. However, the benefit of such a simplification may be questionable, since such an approach would no longer be performance-based, with it, hence, not enabling the most efficient utilisation of the SHCC.

Self-evidently, the limited availability of hard data, meaning data from experimental investigations, is a crucial restriction in the development of a durability framework. Still further, the validity of known findings is negatively influenced by poor design, execution, or analysis of the experimental investigations involved, thus resulting in bias. Other causes of bias include the selective discussion, interpretation, and dissemination of results. In addition, the relevance of published findings must be carefully assessed. For instance, the cracking behaviour of SHCC is typically investigated at an age of between 14 and 28 days, despite its high fly ash content, which is known to significantly alter mechanical properties, including the cracking behaviour, after 28 days have elapsed.

As elucidated by Altmann (2012), the incorporation of expert opinion in durability design concepts is unavoidable if there is insufficient hard data. However, transparency is of utmost importance whenever expert assessment is used. Whether a quantification is based on expert judgement or on hard data must be clearly visible, since, otherwise, there is a significant risk that expert judgement may be disguised as fact. Such a disguise may stymie critical assessment of the quantification and in the long term undermine the trust of users, given the different levels of trust usually afforded to the two sources of information. Furthermore, transparency in the expert reasoning process is of paramount importance, since it is a prerequisite for any meaningful scrutiny and debate regarding expert opinion-based quantification. The required transparency may be achieved through verbal explanations for the variable quantifications. These explanations must not only provide an appropriate context for the quantification, but must also include qualifying statements, as well as reputational evidence. Ideally, fuzzy quantification should be used for such subjective data, as opposed to the stochastic quantification of uncertainty associated with hard data (cf. Sect. 10.4).

Sources of expert opinion and knowledge include:

i. analogies that are drawn from published and unpublished information on, and experience with, ordinary concrete;
ii. interpretations of published and unpublished information on, and experience with, SHCC; and
iii. interpretations of published and unpublished information on, and experience with, other new cementitious materials exhibiting some characteristics that are similar to those of SHCC, such as textile-reinforced concrete.

The above-mentioned sources may be harnessed using the annotated knowledge, judgement, and interpretations of a single expert, or by means of combining the knowledge, judgement, and interpretations of a group of experts.

10.6 Existing Deterioration Models, Design Concepts and Recommendations

10.6.1 Design Recommendations of the Japan Society of Civil Engineers (JSCE)

The only current design recommendations for field applications of SHCC have been published by the Japan Society of Civil Engineers (Rokugo 2008). The recommendations consist of verbal specifications of design principles and a commentary, providing guidance on how to interpret the principles, as well as details of the related Japanese research results that have been achieved so far, so as to assist in the quantification of certain of the input variables involved.

The recommendations cover the following environmental actions:

i. steel corrosion, due to carbonation;
ii. chloride-induced steel corrosion;
iii. freezing and thawing;
iv. chemical attacks;
v. alkali–aggregate reaction; and
vi. fire.

Additionally, the influence of cracking on water tightness is covered.

Only for the initiation period of chloride-induced steel reinforcement corrosion is a limit state approach suggested, with a critical chloride concentration at the reinforcement that must not be exceeded. However, in contrast to the DuraCrete framework, or to the fib Model Code for Service Life Design for RC structures, the recommendations include neither a constitutive equation for chloride ingress, nor variable quantifications, or information, regarding the required level of safety. In principle, adoption of the following step-by-step process is recommended. First, the design tensile strain is to be estimated, based on which the maximum crack width is determined. Its influence on the design diffusion coefficient is computed according to Eq. 10.6 (Rokugo et al. 2007) for cracked SHCC:

$$D_d = D_k + D_{cr} \cdot \log\left(\varepsilon\omega^2\right) \tag{10.6}$$

with:

D_d the design value of chloride diffusivity
D_k a material constant
D_{cr} a constant representing the contribution of cracks and tensile strain to the chloride diffusivity of SHCC
ε the tensile strain
ω the characteristic value of the maximum crack width

Since no constitutive equation is provided, it is not specified, however, how the chloride concentration at the reinforcement is to be determined, based on the available information. This design concentration has, then, to be assessed against the given critical concentration of 1.2 kg/m^3.

Whereas the principles that are set out in the Japanese design recommendations are sound, they do not constitute a comprehensive design concept, since they fail to meet all the requirements for a performance-based design concept that were set out in Sect. 10.3.

10.6.2 Fuzzy-Probabilistic Design Concept

Regarding reinforcement corrosion due to carbonation or chloride ingress, Tuutti (1982) proposes differentiating between an initiation period, during which the deleterious agents make ingress into the concrete, and the subsequent propagation period, once the steel has been depassivated (cf. Fig. 10.7). Ongoing steel corrosion may eventually induce cracking, spalling, and eventual collapse. Depending on the particular local conditions, the corrosion may proceed at very different rates, with the crack pattern and the concrete quality being important parameters.

Following the lead of existing performance-based design concepts for ordinary concrete, Altmann et al. (2012a) defined the end of the initiation period as being the end of service life.

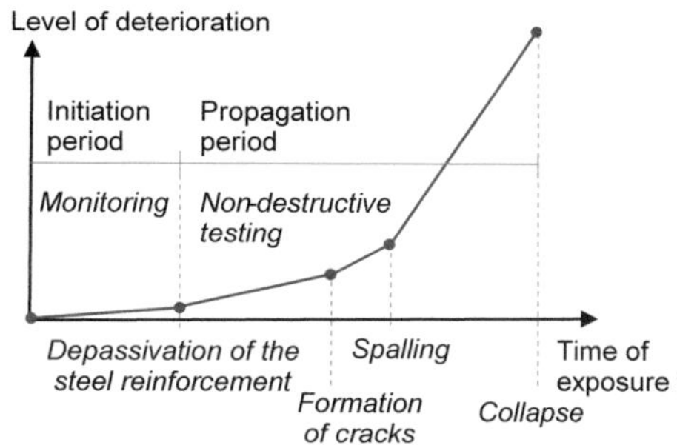

Fig. 10.7 Deterioration process of steel reinforcement corrosion, adapted from Tuutti (1982), and from Gehlen and Kapteina (2004)

Figure 10.8 shows the approach that was followed by Altmann (2012) and Altmann et al. (2012a) for the development of a performance-based fuzzy-probabilistic durability concept for R/SHCC exposed to chlorides. According to the authors concerned, the concept has the potential to be the nucleus of a comprehensive durability design framework for SHCC, and to serve as a template for similar concepts for other new cementitious materials (Altmann and Mechtcherine 2013).

Due to its relative simplicity, flexibility, and transparency, the DuraCrete framework provides an appropriate basis for the new design concept. The empirical deterioration models used in the DuraCrete framework and its derivatives for crack-free ordinary concrete are simple analytical formulas with stochastically

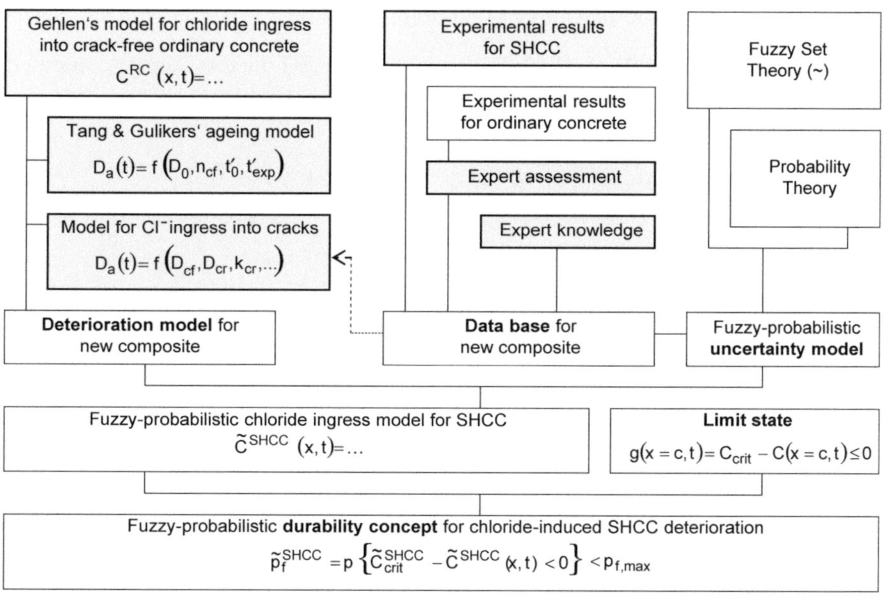

Fig. 10.8 Schematic representation of the development of a performance-based model for chloride ingress into SHCC (Altmann and Mechtcherine 2013)

quantified input variables. Both formulas and input variables may be modified and requantified as required.

Gehlen and Kapteina (2004) propose the acceptance of Eq. 10.7 as a model for chloride ingress into crack-free ordinary concrete:

$$C(x,t) = C_0 + \left(C_{S,\Delta x} - C_0\right) \cdot erfc\left(\frac{x - \Delta x}{2\sqrt{D_a(t) \cdot t}}\right) \tag{10.7}$$

with:

t the duration of exposure
C_0 the initial chloride concentration in the cementitious composite
$C_{S,\Delta x}$ the chloride concentration at the depth of the convection zone Δx

In submerged conditions, $\Delta x = 0$, and $C_{S,\Delta x}$ becomes the surface chloride concentration.

The time-dependent apparent diffusion coefficient $D_a(t)$ for SHCC is given by Eq. 10.8:

$$D_a(t) = D_{ex,0,cf} T_{cf} + D_{cr} T_{cr} \tag{10.8}$$

with:

$D_{ex,0,cf}$ the experimentally determined diffusion coefficient in crack-free SHCC at the reference time t'_0
D_{cr} the contribution of cracks to chloride diffusion

Using Tang and Gulikers' (2007) ageing model, the T-factors can be computed, according to Eqs. 10.9 and 10.10:

$$T_{cf} = k_e k_t k_c \frac{1}{1 - n_{cf}} \left[\left(1 + \frac{t'_{\exp}}{t}\right)^{1-n_{cf}} - \left(\frac{t'_{\exp}}{t}\right)^{1-n_{cf}}\right] \left(\frac{t'_0}{t}\right)^{n_{cf}} \tag{10.9}$$

$$T_{cr} = k_e k_{t,cr} k_{cr} \frac{1}{1 - n_{cr}} \left[\left(1 + \frac{t'_{\exp}}{t}\right)^{1-n_{cr}} - \left(\frac{t'_{\exp}}{t}\right)^{1-n_{cr}}\right] \left(\frac{t'_0}{t}\right)^{n_{cr}} \tag{10.10}$$

with the age factors:

n_{cf} for crack-free SHCC
n_{cr} for the crack contribution to chloride ingress

The *k*-factors account for the differences found between the laboratory and field conditions.

To facilitate the transparent harnessing of all available information, including expert knowledge, the input variables for these equations are quantified using a fuzzy-probabilistic uncertainty model. For the analysis of the design equations, such existing numerical algorithms as the one developed by Sickert (2005) can be utilised. If uncertainty is quantified with fuzzy random input variables, the resulting failure probability is also fuzzy, which means that imprecision due to a lack of available information remains visible. To obtain a crisp, or non-fuzzy, failure probability, the result may be defuzzified, which entails the pursuance of a process for which various options are available, as is described, for example, by Möller and Beer (2004).

The practical applicability of the above design concept must, however, be noted as being somewhat limited. As outlined in Sect. 2.5.1, experimental investigations have shown that, even in the narrow cracks that are present in SHCC, fast ingress of chloride occurs by capillary absorption. Such ingress may result in a corrosion initiation phase (cf. Fig. 10.8) that is of negligible duration when it is compared to a service life of 50 or more years. This means that, if convective transport in cracks cannot be prevented, and the end of service life is defined as being the end of the initiation phase, R/SHCC members will not attain a meaningful service life unless exposure is avoided by means of coating or the resistance against chloride ingress is significantly increased by means of hydrophobising treatment.

In the light of the reduced steel corrosion rate and the higher spalling resistance of SHCC in comparison with ordinary concrete (cf. Sect. 10.1), such findings may also serve as an encouragement to reconsider the appropriateness of the end of initiation phase as a limit state in the durability design of R/SHCC members. A possible limit state would be the loss of the load-bearing capacity of reinforcement, due to the corrosion-induced reduction in rebar cross-sections. However, it should also be kept in mind that chloride ingress, and other types of exposure, may change the mechanical properties of SHCC. This affects the cracking behaviour, and, thus, the level of protection that is afforded the steel. Similar to the case of structural design, different serviceability and ultimate limit states along the x-axis in the Tuutti-based deterioration concept in Fig. 10.7 may be specified. Since the consequences of failure tend to become more severe in line with the progression of further deterioration, the required minimum reliability index β_{min} must be higher the further along the x-axis the specified limit state is located.

10.6.3 Example Application of the Fuzzy-Probabilistic Design Concept

The simplest application example of the above design concept is that of the determination of the time-dependent probability of the failure of a crack-free steel-reinforced SHCC member that is permanently submerged in seawater.

For a specific SHCC composition with experimentally determined chloride diffusivity at different young ages, the derivation of input variables for the computation of the probability of failure is discussed in detail by Altmann et al. (2012b). In the absence of comprehensive information on the issue, a combination of

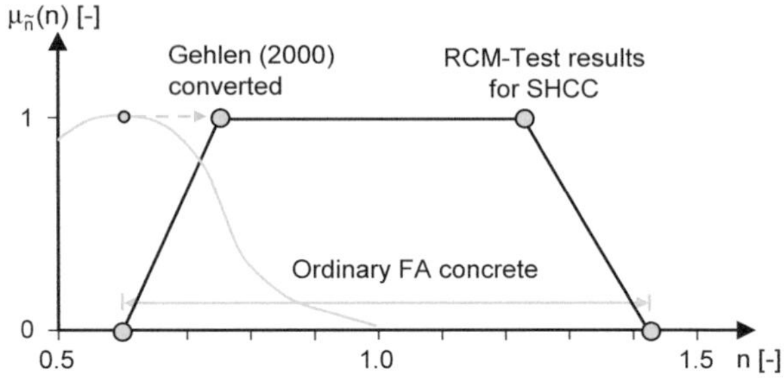

Fig. 10.9 Membership function μ of the fuzzy age factor (Altmann et al. 2012b)

experimental results and of theoretical considerations, as well as analogies with ordinary concrete, as based on expert assessment, were used to determine the variables concerned.

The age factor n, which is used to describe the time-dependence of diffusivity, is a case in point. The factor is usually determined empirically from a large database consisting of a material's diffusivity after both short and long periods of exposure. Altmann et al. (2012b) quantified the parameter concerned, based on a limited number of short-term laboratory experiments, and on expert assessment. The experimental investigations yielded a value of 1.23 for the age factor. For ordinary fly ash concrete and for a somewhat different equation by means of which to determine the apparent diffusivity, typical values in the literature range from 0.60 to 0.70, with outliers being as low as 0.48 and as high as 0.91. The authors deem it reasonable to assume that the above-mentioned outliers are also possible realisations for the age factor of SHCC. However, they consider the possibility that it shows values that are typical for ordinary fly ash concrete, or that are determined for the specific SHCC composition to be significantly higher. Thus, after conducting a conversion to account for the different formula for the apparent diffusivity, the age factor was quantified with the fuzzy uncertainty according to Fig. 10.9.

For a critical chloride concentration according to the fib's (2006) Model Code for Service Life Design, the time-dependent probability of failure given in Fig. 10.10 was computed by the authors concerned. The pronounced non-stochastic uncertainty of the computed failure probability might suggest that the proposed durability concept does not, as yet, yield meaningful results. In fact, the opposite is true. Even with a very limited database, it could be shown that a transparent determination of the probability of failure is possible. Still further, a sensitivity analysis of the result revealed that a reduction of the non-stochastic uncertainties associated with only one key input variable is likely to result in a significant reduction of the non-stochastic uncertainty. In such a case, even the most conservative defuzzification would result in a crisp failure probability that would increase sufficiently slowly over time to result in a meaningful service life (Altmann 2012; Altmann et al. 2012b).

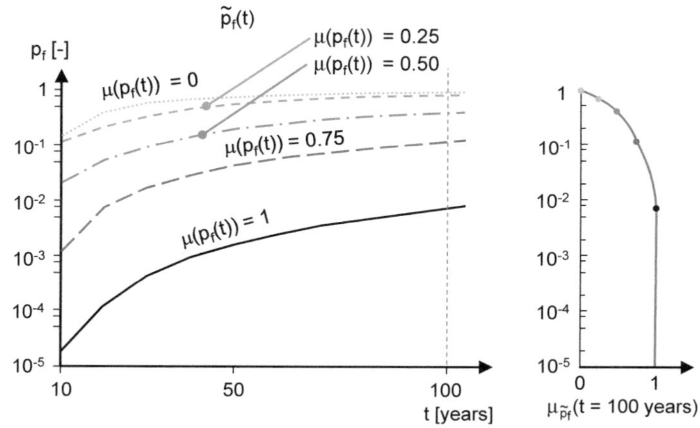

Fig. 10.10 Membership function μ of the fuzzy probability of failure as a function of time (*left*) and after 100 years of exposure (*right*) (Altmann et al. 2012b)

10.6.4 Deterioration Models

As has been discussed in the previous chapters of the present text, deterioration models, or elements thereof, exist for some load cases other than chloride ingress into SHCC.

In Sect. 1.6, the current state-of-the-art regarding the modelling of crack pattern development is presented. As was noted in Chap. 1, the scientific models for the crack opening at maximum fibre resistance do not predict the crack width at typical serviceability strain levels, thus making them unsuitable for service life design. Further to the above, model-free stochastic crack width distributions are reported in the literature. As for crack spacing, a purely empirical model, without clear physical meaning of the input variables involved, is provided.

At present, neither empirical nor scientific models exist for the quantification of crack patterns — that is, for crack width and crack spacing as a function of composition, load, time, and additional bar reinforcement. Hence, an a priori modelling of cracking as a deterioration process in the context of service life or durability design is not yet possible.

In Chap. 2, linking the transport of deleterious media in and through SHCC to the crack pattern is proposed. To such an end, empirical models have been developed. Firstly, the so-called Ingress Potential Index (IPI) is a function of (i) the deleterious medium (i.e. water and chlorides), and (ii) the crack pattern. However, as was noted in Sect. 2.2, "currently, the concept (based on the IPI Index) does not allow for the quantitative derivation of a flow rate, or of an absorbed water volume, for a certain crack pattern under a given exposure". Secondly, the hydraulic crack pattern parameter describes water transport through cracked SHCC based on actual transport models as a function of the crack pattern. Thirdly, for capillary absorption, an empirical function of the absorbed water volume has been developed that considers the capillary absorption of the crack-free bulk material and the contribution of the cracks. For none of the transport models are parameter quantifications presented for an a priori member/structure durability design that meets the requirements as set out in Sect. 10.3.1. Thus, the quantification of the relationship between transport processes and structural durability is an open issue that requires intense focus.

For fibres, no deterioration models are given in Chap. 3. Chaps. 4 to 7 provide no deterioration models for matrix and fibre–matrix interface degradation.

For the bond between SHCC overlays and the concrete substrate, an empirical model is presented in Chap. 8. As is outlined there, the model was recently further developed into what can best be described as a scientific model for the mechanical interface behaviour of SHCC overlays. At this stage, it is highly likely that the input variables for the model concerned exhibit similar uncertainty as do those for the chloride ingress model described above.

Finally, Chap. 9 presents an empirical model for the steel rebar corrosion depth in cracked R/SHCC exposed to chlorides. However, additional work is required to quantify input variables so as to allow for an a priori durability design. No deterioration models for other load cases are provided.

In summary, it can be concluded that, only for crack pattern development, chloride ingress, and SHCC overlay bond behaviour degradation models exist. Significant research efforts are still required to quantify the respective input variables with a reasonably low uncertainty, so as to allow for meaningful service life design.

10.7 Summary and Outlook

10.7.1 Summary

To fully utilise the advantageous durability properties of SHCC for the efficient design of durable members and structures, a comprehensive framework for performance-based durability or service life design is required. This framework must be capable of encompassing all relevant load cases and load case combinations.

A first step in this direction has already been taken with the development of a fuzzy-probabilistic durability concept for chloride-induced rebar corrosion in R/SHCC members. However, to develop the concept into a comprehensive durability design framework, the uncertainty of its input variables must be reduced, its deterioration model must be further validated, and the concept must be expanded to include other limit states, load cases, and load case combinations.

From Fig. 10.1 and the small number of models for relevant deterioration processes presented in this chapter it is obvious that the development of a comprehensive design framework presents a significant challenge. In light of the slow progress made in the development of a similar framework for ordinary concrete, for which an abundance of information is available, it is obvious that, for most load cases and load case combinations, no empirical or scientific deterioration models with reasonably well-quantified input variables are likely to become available in the foreseeable future.

As discussed in Sect. 10.5.3, this means that causal reasoning of the 'if-then' type, such as "*if the fly ash content of a composition increases, the decrease in diffusivity over time increases as well*", using all available hard data and subjective information is likely to be the only option to describe most deterioration processes and assess the durability of (R/)SHCC structures and members.

10.7.2 Deterioration Model-Free Durability Assessment with Fuzzy Uncertainty Quantifications

Altmann (2012) and Altmann and Mechtcherine (2013) propose the use of fuzzy 'if–then' rules and fuzzy cognitive maps (FCMs), as outlined below, in lieu of non-existent deterioration models.

In the presence of nonstochastic uncertainty, 'if–then' rules become fuzzy. Where expert judgement is harnessed, this fuzziness often assumes the form of vague linguistic assessments such as '*a lot*' or '*a little*'. Figure 10.11 shows a

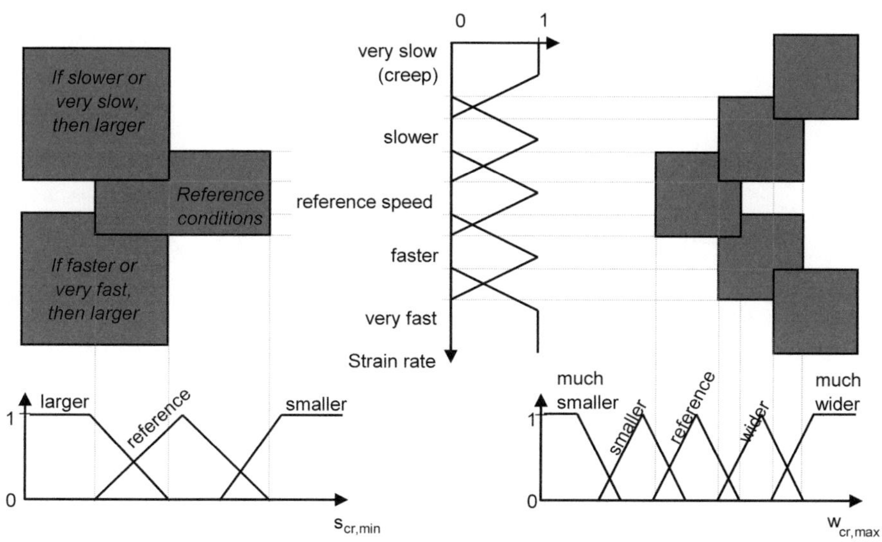

Fig. 10.11 Fuzzy rules describing maximum crack width $w_{cr,max}$ and minimum crack spacing $s_{cr,min}$ as a function of the tensile strain rate (Altmann 2012)

visualisation of such fuzzy rules for the crack status as a function of the strain rate. Fuzzy rules have the advantage of being highly transparent and flexible, and with numerous such rules firing simultaneously to different degrees, they allow model-free forecasting (Kosko 1993).

Fuzzy rules may be combined to form fuzzy cognitive maps (FCMs), which can be defined as "fuzzy-graph structures for representing causal reasoning" (Kosko 1986), and which allow the extraction, analysis, and visualisation of knowledge based on hard evidence, as well as subjective expert judgement. In contrast to tree-type mapping, FCMs allow for backward chaining. FCMs and their analysis can support a statement and make its reasoning transparent but cannot prove it (Kosko 1993).

Figure 10.12 shows a simple cognitive map of the influence of fly ash on chloride ingress. Each arrow defines a causal link or connection, with a *plus* signifying causal increase and a *minus* signifying causal decrease. For instance, if the fly ash content is increased, the *pH* in the composite decreases, and if chloride binding increases, so does the critical (total) chloride concentration C_{crit}. Being fuzzy, these rules can be weighted with any number between 0 and 1, as well as linguistic assessments such as '*a little*' or '*significantly*'.

With FCMs and freely available software like the FCMapper (Bachhofer 2009), knowledge about an individual component such as fly ash, an individual composition, or an individual load case such as chloride ingress may be described. Based on the available knowledge, they allow for the assessment of the consequences of changed boundary conditions, such as the material composition for chloride ingress, based on expert rules.

Knowledge can also be grown by combining different FCMs to the point shown schematically in Fig. 10.12b, where they include various load cases and their interactions for various compositions. This is especially interesting for the

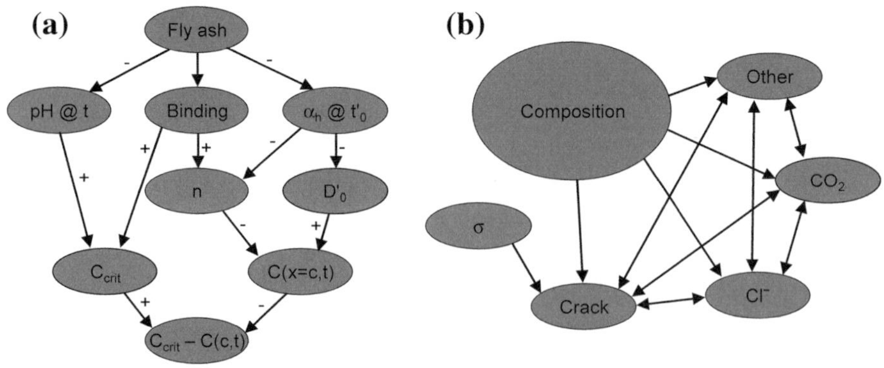

Fig. 10.12 **a** Cognitive map of the influence of the fly ash content on the initiation of chloride-induced reinforcement corrosion in ordinary concrete. *Positive arrows* represent causal increases, *negative* ones causal decreases. **b** Cognitive map of the degradation of SHCC and R/SHCC members (Altmann 2012)

assessment of load combinations, for which it is unlikely that comprehensive information will become available in the foreseeable future.

FCMs are powerful tools for transparently quantifying subjective information and assessing deterioration processes based on information that does not require specialist modelling knowledge and experience. They can easily be amended and expanded on if and when any additional information becomes available. It follows, that FCMs are very well suited as the basis for cooperation between researchers focusing on (experimental) investigations and those focusing on model development. Hence, it is hoped that they will streamline and expedite the development of deterioration models with input parameter quantifications suitable for the durability design of SHCC members and structures.

References

Alexander, M.G., Ballim, Y., Standish, K. (2008). A framework for use of durability indexes in performance-based design and specifications for reinforced concrete structures. Materials and Structures 41:921-936.

Altmann, F. (2010). Dauerhaftigkeitskonzept für neuartige zementgebundene Werkstoffe am Beispiel hochduktilen Betons (SHCC). Breit, W., Kurz, W., Schnell, J., Kohlmeyer, C. (eds.), Beiträge zum 51. Forschungskolloquium des DAfStb, Kaiserslautern, Technische Universität Kaiserslautern, pp. 697-708.

Altmann, F. (2012). A durability concept for strain-hardening cement-based composites. Doctoral thesis, Technische Universität Dresden, Germany.

Altmann, F., Sickert, J.-U., Mechtcherine, V., Kaliske, M. (2012a). A fuzzy-probabilistic durability concept for strain-hardening cement-based composites (SHCCs) exposed to chlorides: Part 1: Concept development. Cement and Concrete Composites 34:754-762.

Altmann, F., Sickert, J.-U., Mechtcherine, V., Kaliske, M. (2012b). A fuzzy-probabilistic durability concept for strain-hardening cement-based composites (SHCC) exposed to chlorides: Part 2: Application example. Cement and Concrete Composites 34:763-770.

Altmann, M., Mechtcherine, V. (2013). Durability design strategies for new cementitious materials. Cement and Concrete Research 54:114-125.

Bachhofer, M. (2009). FCMapper, www.fcmappers.net.

Baroghel-Bouny, V., Kinomura, K., Thiery, M., Moscardelli, S. (2011). Easy assessment of durability indicators for service life prediction or quality control of concretes with high volumes of supplementary cementitious materials. Cement and Concrete Composites 33 (8):832-847.

Buckley, J.J. (2005). Fuzzy probabilities. Springer.

CUR (2009). Durability of structural concrete with regards to chloride induced reinforcement corrosion - Guideline for formulating performance requirements (in Dutch). CUR, Gouda.

DAfStb – Deutscher Ausschuss für Stahlbeton (2008). Positionspapier des DAfStb zur Umsetzung des Konzepts von leistungsbezogenen Entwurfsverfahren unter Berücksichtigung von DIN EN 206-1. Anhang J. Beton- und Stahlbetonbau 103(12):837-839.

DIN EN 206-1:2001-07 (2001). Concrete – Part 1: Specification, performance, production and conformity; German version EN 206-1:2000.

DIN EN 1990:2010-12 (2010). Eurocode: Basis of structural design; German version EN 1990:2002 + A1:2005 + A1:2005/AC:2010.

DIN EN 1992-1-1:2011-01 (2011). Eurocode 2: Design of concrete structures – Part 1-1: General rules and rules for buildings; German version EN 1992-1-1:2004 + AC:2010.

DuraCrete (2000a). Statistical quantification of the variables in the limit state function. The European Union – Brite Euram III, Project BE95-1347, Report R9.

DuraCrete (2000b). Probabilistic performance based durability design of concrete structures. The European Union – Brite Euram III, Project BE95-1347, Report R15.

DuraCrete (2000c). Final technical report. The European Union – Brite Euram III, Project BE95-1347, Report R17.

Edvardsen, C. (2010). The consultants view on service life design. Proceedings of the Conference on Service Life Design for Infrastructure (PRO 70), Delft, RILEM Publications S.A.R.L., pp. 249-264.

Gehlen, C. (2000). Probabilistische Lebensdauerbemessung von Stahlbetonbauwerken. Schriftenreihe des Deutschen Ausschusses für Stahlbeton Heft 510, Beuth.

Gehlen, C., Kapteina, G. (2004). DARTS – Durable and Reliable Tunnel Structures, Deterioration modelling. The European Union – GROWTH 2000, Project GRD1-25633.

Gehlen, C., Schießl, P., Schießl-Pecka, A. (2008). Hintergrundinformationen zum Positionspapier des DAfStb zur Umsetzung des Konzepts von leistungsbezogenen Entwurfsverfahren unter Berücksichtigung von DIN EN 206-1, Anhang J, für dauerhaftigkeitsrelevante Problemstellungen. Beton und Stahlbetonbau 103(12):840-851.

Gehlen, C., Von Greve-Dierfeld, S. (2010). Durability of reinforced concrete structures: Design formats. Proceedings of the Conference on Service Life Design for Infrastructure (PRO 70), Delft, RILEM Publications S.A.R.L., pp. 351-358.

Gulikers, J. (2007). A critical review of the mathematical modelling of chloride ingress into concrete and the derivation of input data. Proceedings of the International RILEM Workshop on Performance Based Evaluation and Indicators for Concrete Durability (PRO 47), RILEM Publications S.A.R.L., pp. 165-175.

Gulikers, J. (2011). Experience with a performance-based approach at Rijkswaterstaat. Presentation at fib TG 8.10 Meeting, Leipzig.

fib – International Federation for Structural Concrete (2006). Model code for service life design. fib Bulletin 34, Lausanne.

fib – International Federation for Structural Concrete (2010). Model code 2010 – first complete draft, Volume 1. fib Bulletin 55, Lausanne.

ISO 16204:2012 (2012). Durability – Service life design of concrete structures. International Organization for Standardization.

JCSS – Joint Committee on Structural Safety (2001a). Probabilistic model code – Part 1: Basis of design.

JCSS – Joint Committee on Structural Safety (2001b). Probabilistic model code – Part 2: Load models.

JCSS – Joint Committee on Structural Safety (2001c). Probabilistic model code – Part 3: Material properties.

Kosko, B. (1993). Fuzzy thinking: The new science of fuzzy logic. Hyperion, New York.

Kosko, B. (1986). Fuzzy cognitive maps. International Journal of Man-Machine Studies 24:65-75.

Kunieda, M., Rokugo, K. (2006). Recent progress on HPFRCC in Japan. Journal of Advanced Concrete Technology 4(1):19-33.

Lay, S., Schießl, P. (2003). LIFECON – Life Cycle Management of Concrete Infrastructure for Improved Sustainability, LIFECON Deliverable D 3.2 – Service life models. The European Union – GROWTH 2000, Project G1RD-CT-2000-00378.

Lee, Y.-K., Hwang, D.-S. (2008). A study of the techniques of estimating the probability of failure. Journal of the Chungcheong Mathematical Society 21(4):573-583.

Lepech, M.D., Li, V.C. (2009). Application of ECC for bridge deck link slabs. Materials and Structures 42:1185-1195.

Li, V.C. (2003). On Engineered Cementitious Composites (ECC): A review of the material and its applications. Journal of Advanced Concrete Technology 1(3):215-230.

Mechtcherine, V., Altmann, F. (2011). Durability of structural elements and structures. Van Zijl, G.P.A.G., Wittmann, F.H. (eds.), Durability of strain-hardening fibre-reinforced cement-based

composites (SHCC). State-of-the-Art Report prepared by Subcommittee 2 of RILEM Technical Committee 208-HFC, Springer, pp. 89-112.
Mechtcherine, V. (2012). Towards a durability framework for structural elements and structures made of or strengthened with high-performance fibre-reinforced composites. Construction and Building Materials 31:94-104.
Mechtcherine, V. (2013). Novel cement-based composites for the strengthening and repair of concrete structures. Construction and Building Materials 41:365-373.
Melchers, R.E. (1987). Structural reliability – Analysis and prediction. Ellis Horwood, Chichester.
Möller, B., Graf, W., Beer, M. (2003). Safety assessment of structures in view of fuzzy randomness. Computers & Structures 81(15):1567-1582.
Möller, B., Beer, M. (2004). Fuzzy randomness. Springer.
Müller, H.S., Vogel, M. (2008). Lebenszyklusmanagement im Betonbau. beton 58(5):206-215.
Nilsson, L.-O. (2005). CHLORTEST, WP4-Report – Modelling of chloride ingress. The European Union – GROWTH 2000, Project GRD1-2002-71808.
Reuter, U. (2006). Analyse und Prognose von Zeitreihen mit Fuzzy-Daten zur Prädiktion von Strukturantworten. Veröffentlichungen des Instituts für Statik und Dynamik der Tragwerke, Heft 10, Technische Universität Dresden, Germany.
Rokugo, K., Kanda, T., Yokota, H., Sakata, N. (2007). Outline of JSCE recommendation for design and construction of multiple fine cracking type fiber reinforced cementitious composite (HPFRCC). H.W. Reinhardt, A.E. Naaman (eds.), Proceedings of the 5th International Conference on High Performance Fiber Reinforced Composites (HPFRCC5), 10-13 July 2007, Mainz, Germany, PRO 53, RILEM Publication S.A.R.L., pp. 203-212.
Rokugo, K. (ed.) (2008). Recommendations for design and construction of high performance fiber reinforced cement composites with multiple fine cracks (HPFRCC). Concrete Engineering Series 82. Japan Society of Civil Engineers (JSCE).
Rokugo, K., Kanda, T., Yokota, H., Sakata, N. (2009). Applications and recommendations of high performance fiber reinforced cement composites with multiple fine cracking (HPFRCC) in Japan. Materials and Structures 42(9):1197-1208.
Rokugo, K., Kanda, T. (eds.) (2013). Strain hardening cement composites: Structural design and performance. State-of-the-Art Report prepared by Subcommittee 3 of RILEM Technical Committee 208-HFC, Springer.
Schuëller, G.I. (1997). A state-of-the-art report on computational stochastic mechanics. Probabilistic Engineering Mechanics 12(4):197-321.
Schuëller, G.I., Pradlwarter, H.J., Koutsourelakis, P.S. (2003). A comparative study of reliability estimation procedures for high dimensions. Proceedings of the 16th ASCE Engineering Mechanics Conference, 16-18 July 2003, Seattle, University of Washington.
Sickert, J.-U. (2005). Fuzzy-Zufallsfunktionen und ihre Anwendung bei der Tragwerksanalyse und Sicherheitsbeurteilung. Veröffentlichungen des Instituts für Statik und Dynamik der Tragwerke, Heft 9, Technische Universität Dresden, Germany.
Söderqvist, M.-K., Vesikari, E. (2004). LIFECON – Life Cycle Management of Concrete Infrastructure for Improved Sustainability, LIFECON Deliverable D 1.1 – Generic technical handbook for a predictive life cycle management system of concrete structures (LMS). The European Union – GROWTH 2000, Project G1RD-CT-2000-00378.
Tang, L., Gulikers, J. (2007). On the mathematics of time-dependent apparent chloride diffusion coefficient in concrete. Cement and Concrete Research 37(4):589-595.
Tuutti, K. (1982). Corrosion of steel in concrete. CBI forskning/research Fo 4:82, Swedish Cement and Concrete Research Institute, Stockholm.
U.S. Department of Commerce (1991). Fuzzy logic: A key technology for future competitiveness. National Technical Information Service.
Van Zijl, G.P.A.G., Wittmann, F.H. (eds.) (2011). Durability of strain-hardening fibre-reinforced cement-based composites (SHCC). State-of-the-Art Report prepared by Subcommittee 2 of RILEM Technical Committee 208-HFC, Springer.
Van Zijl, G.P.A.G., Wittmann, F.H., Oh, B.H., Kabele, P., Toledo Filho, R.D., Fairbairn, E.M.R., Slowik, V., Ogawa, A., Hoshiro, H., Mechtcherine, V., Altmann, F., Lepech, M.D. (2012).

Durability of strain-hardening cement-based composites (SHCC). Materials and Structures 45 (10):1447-1463.

Viertl, R., Hareter, D. (2006). Beschreibung und Analyse unscharfer Informationen – Statistische Methoden für unscharfe Daten. Springer.

Wagner, C., Slowik, V., Waldenburger, K. (2008). Dehnungsverfestigendes zementgebundenes Material für die Sanierung gerissener Betonflächen. Bautechnik 85(1):49-56.

Wittmann, F.H. (2011). Durability under combined loads. Van Zijl, G.P.A.G., Wittmann, F.H. (eds.), Durability of strain-hardening fibre-reinforced cement-based composites (SHCC). State-of-the-Art Report prepared by Subcommittee 2 of RILEM Technical Committee 208-HFC, Springer, pp. 73-79.

Yamamoto, T., Nagoya, K., Hiraishi, T., Sakata, N., Shimizu, M., Mashimoto, H. (2005). Tunnel repair technology using highly ductile cement composites - An application to a tunnel damaged during Chuetsu Earthquake (in Japanese). JSCE Annual Journal of Materials, Concrete Structures and Pavements 60(3):483-484.

MIX
Papier aus verantwortungsvollen Quellen
Paper from responsible sources
FSC® C105338

If you have any concerns about our products,
you can contact us on
ProductSafety@springernature.com

In case Publisher is established outside the EU,
the EU authorized representative is:
Springer Nature Customer Service Center GmbH
Europaplatz 3, 69115 Heidelberg, Germany

Printed by Libri Plureos GmbH
in Hamburg, Germany